ANIMAL BEHAVIOR: NEW RESEARCH

ANIMAL BEHAVIOR: NEW RESEARCH

EMILIE A. WEBER AND LARA H. KRAUSE
EDITORS

Nova Science Publishers, Inc.
New York

LIBRARY OF CONGRESS CATALOGING-IN-PUBLICATION DATA

Animal behavior : new research / Emilie A. Weber and Lara H. Krause (editors).
   p. cm.
 ISBN 978-1-60456-782-3 (hardcover)
 1. Animal behavior--Research. I. Weber, Emilie A. II. Krause, Lara H.
 QL751.A6499 2008
 591.5--dc22
               2008020943

Published by Nova Science Publishers, Inc. New York

CONTENTS

Preface   **vii**

Chapter 1   Sexual Selection, Mate Choice, and Primate Preferences   **1**
            Jack A. F. Griffey and Anthony C. Little

Chapter 2   Female Mate Choice in Non-human Mammals   **35**
            Benjamin D. Charlton

Chapter 3   Partition Test and Sexual Motivation in Male Mice   **57**
            N. N. Kudryavtseva

Chapter 4   Gaze Following in Non-human Animals:
            The Corvid Example   **73**
            Christian Schloegl, Judith Schmidt,Christelle Scheid,
            Kurt Kotrschal and Thomas Bugnyar

Chapter 5   How Different Host Species Influence Parasitism
            Patterns and Larval Competition of Acoustically-
            Orienting Parasitoid Flies (Tachinidae: Ormiini)   **93**
            Gerlind U. C. Lehmann

Chapter 6   Insights into the Acoustic Behaviour of Polar Pinnnipeds –
            Current Knowledge and Emerging Techniques of Study   **133**
            Ilse Van Opzeeland, Lars Kindermann,
            Olaf Boebel and Sofie Van Parijs

Chapter 7   Food Hoarding in the New Zealand Robin:
            A Review and Synthesis   **163**
            Ignatius J. Menzies and K. C. Burns

Chapter 8   Discriminative Learning, Learning Generalization
            and Masking Tests as Three Strategies to
            Assess Olfactory Discrimination   **185**
            Julien Colomb

Chapter 9   Constraints and the Evolution of Mutual Ornamentation   **193**
                Ken Kraaijeveld and Barbara M. Reumer

Index   **215**

PREFACE

This new book is devoted to recent research on animal behavior which relates to what an animal does and why it does it. The types of behaviors exhibited are rich and various. Some are genetically determined, or instinctive, while others are learned behaviors. The desire to understand the animal world has made ethology a rapidly growing field, and since the turn of the 21st century, many prior understandings related to diverse fields such as animal communication, personal symbolic name use, animal emotions, animal culture and learning, and even sexual conduct, long thought to be well understood, have been revolutionized.

Chapter 1 - Mate choice and the preferences that many animals display when selecting potential mates are, as Kokko *et al.* (2003) explain, incredibly important evolutionary processes which impose selection upon the opposite sex and are accountable for a vast array of spectacular ornaments and characteristics that remain inexplicable via natural selection alone. Consequently, due to the evolutionary importance of mate choice and perhaps, as Bateson (1983) suggests, the renewed vitality of areas such as evolutionary and population biology, since the 1970s sexual selection and its implications for mate choice have experienced a rapid revival in interest. Based upon a number of major theoretical insights and empirical findings there has been a growing interest in the mating preferences of animals to the point where mate choice and sexual selection have been two of the most active disciplines of scientific research within behavioural ecology and evolutionary biology.

Chapter 2 - Until now definite studies of mate choice have typically focused on non-mammal animal species where short life spans and gestation periods make reproductive success quicker to determine, and in which the females show preferences for clearly defined morphological or behavioural male traits. Mammals, which are invariably larger, have longer life spans and inter-birth intervals and are behaviourally more complex, are less suited to this type of experimentation. However, notwithstanding these difficulties, many studies of mate choice have been conducted on mammals and the current literature reveals that the females of several mammal species do appear to choose their mates, and through this potentially gain important direct and indirect fitness benefits. Here the author reviews this body of work to reveal that the majority of female mammal mate choice studies conducted so far, particularly on large mammals, fail to: 1) determine the actual male phenotypic trait (s) of female preference (which is crucial to identifying male characteristics under sexual selection); 2) Take into account other environmental and social factors that may affect female response to male roars, as well as intrinsic factors such as female hormonal state and breeding status; and 3) quantify the fitness benefits to discriminating females. The author goes on to give suggestions for future research and emphasise the need for a combination of carefully designed experimental and field studies. Experimental setups allow us isolate specific male traits from other aspects of the male phenotype, determine female hormonal state and control for other competing male and female mating strategies that may affect female behaviour. Field observations of female mating behaviour can then be conducted to determine whether behavioural responses reported using experimental setups translate to actual copulations in natural conditions, and hence affect the reproductive success of individuals. Only this integrative approach will allow us to gain an appreciation of how inter-sexual selection is generated on specific aspects of the male phenotype, and ultimately enable us to understand the link between mating preferences, female mate choice and reproductive success in mammals.

Chapter 3 - Theoretical analysis of own and literature investigations of sexual motivation with the use of the partition test [Kudryavtseva, 1987, 1994] in male mice was carried out. It has been shown that appearance of a receptive female in the neighboring compartment of common cage separated by perforated transparent partition produces the enhancement of testosterone level in blood and stimulates the behavioral activity near partition as a reaction to the receptive female in naive males. In many studies this behavioral activity is considered as sexual motivation, arising in this experimental context in male mice. The lack of correlation between behavioral parameters and gonad reaction of males on receptive female, uninterconnected changes of these two parameters as well as the lack of sexual behavior between naive male and female when partition is removed cast doubt on this data interpretation. It has been supposed that in naive males behavioral reaction to a receptive female is induced by positive incentive – odor of the female associated with nursing and warmth from mother and other females which look after posterity. Short-term increase of the level of testosterone (possessing rewarding properties) is innate stimulus-response reaction which stimulates and prolongs behavioral interest of male to receptive female. It has been supposed that after sexual experience female odor is associated in experienced males with sexual behavior directed to the sexual partner and resulted in the formation of sexual motivation. The data are considered also in the light of the theory of motivated behavior including "liking", "wanting" and "learning" [Robinson and Berridge, 1993, 2000].

Chapter 4 - By definition, following others' gaze occurs, if an individual shifts its attention towards the visual focus of others. Originally considered to be an unique human skill, this view has been challenged by pioneering work of Povinelli & Eddy (1996), who showed gaze following abilities in chimpanzees (*Pan troglodytes*). Since then, many mammalian species have been found to use others' gaze as a source of information with apes being capable to follow gaze geometrically behind visual barriers. This suggests that they may even understand how barriers impair their own visual perception. Taken together, the data suggest that the use of gaze cues may represent an important skill to obtain information in social mammals. However, hardly anything is known about such skills in non-mammalian species, rendering the evolutionary history of gaze following unclear.

With respect to birds,the authors' knowledge is limited to a single species, the common raven (*Corvus corax*). In order to gain an evolutionary perspective, the authors here present data on two closely related corvids, rooks (*C. frugilegus*) and jackdaws (*C. monedula*). As ravens, the food-caching rooks appear to be capable of following others' gaze behind barriers; additionally, the temporal pattern of the ontogenetic development of gaze following is similar in both species. In contrast, the authors found no evidence that the non-caching jackdaws follow human gaze, but have some indication that they may follow conspecifics' gaze. An understanding of barriers, however, could not be demonstrated in jackdaws. These findings support the idea that geometrical gaze following and gaze following into distant space might serve different functions. The authors suggest that gaze following into distant space may primarily reflect an anti-predator response, whereas geometrical gaze following skills may have developed to protect food caches from scroungers. Finally, the authors will discuss alternative explanations and will present an outline for potential future research.

Chapter 5 - Sexual signals are often critical for mate attraction and reproduction but their conspicuousness can expose the signallers to parasites and predators. In orthopteran insects males typically produce acoustic signals to attract females for mating. Tachinid flies of the family Ormiini act as illicit receivers by detecting the mating songs of their bushcricket and gryllid hosts. Ormiini flies can be characterised as opportunistic hunters; the taxonomic specificity of these flies is moderately low and they can have a range of alternative hosts.

This chapter is in two parts. In the first the author quantify the influence host species has on body size and life history traits of different populations of *Therobia leonidei*, the only European Ormiini fly species. For parasitoids host size can constrain offspring growth, subsequently influencing the evolution of body size and life history traits. The author compared fly populations developing in two hosts: *Poecilimon mariannae* and in the lighter *P. thessalicus*. The fly populations investigated had no substantial morphological or molecular-genetic differences. The general pattern of larval competition was similar in both hosts; increasing parasitoid brood size reduced pupal weight and survival to adulthood. Consistent with a local adaptation hypothesis, pupal weight in the heavier host was about 30% heavier than are those parasitizing the lighter host. Similarly, the critical weight necessary for successful hatching of fly maggots was significantly lower in the lighter host. In contrast, brood size was similar between host species.

In the second part the author reviews studies of gryllid and tettigoniid parasitizing Ormiini flies. The general pattern of host usage for temperate species is quite similar, with pupae from single infected hosts weighing around 10 percent of host weight. One remarkable exception is the Australian *Homotrixa alleni*, a very large fly that parasitizes extremely heavy bushcricket species. In this fly species single pupa weigh less than four percent of host weight, leading to higher mean parasitoid clutch sizes. Coupled to this reduced parasitoid-to-host weight ratio is prolonged host survival and an increase in the probability of superparasitism.

Chapter 6 - This chapter will provide a review of the acoustic behaviour of polar pinnipeds. It will also present a detailed update of new and emerging passive acoustic technologies and how these can further the study of behaviour for polar marine mammals.

Both Arctic and Antarctic pinnipeds are known to exhibit a range of adaptations which enable them to survive and reproduce in an ice-dominated environment. However, large gaps still exist in our understanding of the fundamental ecology of these species, as investigations are severely hampered by the animals' inaccessibility. Improving our understanding of ice-breeding species and the effects that changes in habitat might have on their behaviour is vital, as current climatic trends are rapidly altering the polar environments.

For pinnipeds, acoustic communication is known to play an important role in various aspects of their behaviour. Mother-pup reunions and the establishment of underwater territories during the mating season are examples which, for the majority of species, are known to be mediated by vocal signalling. Acoustic measurements therefore provide an essential tool to study ice-breeding pinnipeds as recordings can be used to remotely monitor sounds, track animal movements and determine seasonal changes in movements and distribution. Recent advances in recording technologies, now allow the acquisition of continuous long-term acoustic data sets, even from the remotest of polar regions.

To date, a range of different types of passive acoustic instruments are used, the choice of which depends largely on the purpose of the study. These instruments in addition to computer-based methods that have been developed for automated detection, classification and localization of marine mammal sounds will be discussed. The autonomous PerenniAL Acoustic Observatory in the Antarctic Ocean (PALAOA) is presented here as one example of such recording systems.

Chapter 7 - Experiments on avian food hoarding have played a prominent role in our understanding of animal behaviour. Important insights into cooperative breeding, foraging theory, social learning, spatial memory and territoriality have been generated by studies investigating how birds create, protect and retrieve caches. However, understanding avian food hoarding is limited to just a few groups of birds, namely acorn woodpeckers, shrikes, parids (tits and chickadees) and corvids (crows, jays and ravens). Given the breadth of information gained from a small number of model organisms, studies on different food hoarding birds may generate valuable new insights into animal behaviour. The New Zealand robin's food hoarding behaviour is unusual and poorly understood. Robins hunt some of the world's largest invertebrate prey, which they cannot consume whole. Large prey are instead dismembered and stored piecemeal in tree canopies. Like many other birds endemic to isolated islands, robins lack pronounced anti-predatory behaviours and are fearless of humans. Wild birds readily cache prey offered to them by hand and recent field experiments have shed light on some aspects of their food hoarding behaviour. Here, the authors synthesize current understanding in a review of food hoarding by the New Zealand robin. They begin with a brief review of robin life history, emphasizing the importance of food hoarding. Next the authors discuss the results of recent field experiments investigating the ecological and social factors regulating when, where and how robins hoard food. Lastly the authors discuss the uniqueness of New Zealand robins by emphasizing differences with other food-hoarding systems and how these differences may have arisen from New Zealand's unique ecological and evolutionary history.

Chapter 8 - While understanding olfactory information coding in the brain is rapidly rising, knowledge of how different odors are perceived and discriminated in the brain remains sparse and speculative. One of the primary reasons for this is the difficulty in assessing if an animal can discriminate odor A from odor B. The author reviews here three different strategies envisaged so far and their caveats: discriminative learning, learning generalization and masking tests.

Discriminative learning involves presenting an odor A with a reinforcer and B alone. The animal is then tested for its preference for A over B. The result depends both on the learning and memory abilities of the animal and on the similarity of the odor used (animals cannot perform well if the A and B are too similar). Generalization learning involves presenting only A either with the reinforcer (associative learning generalization) or without (sensitivity to A will decrease, and cross-adaptation can be tested). By testing the behavioral response toward B, the generalization from A to B is measured. The magnitude of this generalization was shown to be dependent on the similarity of A and B. These paradigms have been used with some success, although their dependence on learning tasks complicates interpretations. A third strategy, called a masking test, has not been extensively used so far, despite its simplicity and efficacy. Here, the chemotaxis of animals toward A is measured in the background of a saturated concentration of B. Although the biological mechanisms involved in this test are debated (olfactory adaptation to B was postulated to be crucial), it seems independent of learning mechanisms and may thus help to dissect the molecular and cellular bases of olfactory discrimination in the future.

Chapter 9 - Elaborate ornamental traits occur in both males and females of many species. With the accumulation of evidence supporting an adaptive role for female ornaments, the role of genetic and physiological constraints is largely ignored. Here, the authors investigate phylogenetic patterns of male and female ornamentation in nine disparate animal taxa. In two cases the authors find evidence consistent with mutual ornamentation resulting from genetic constraints as proposed by Lande (1980), in which sexually monomorphic ornamentation is a temporary stage in the evolution towards sexual dimorphism. However, the authors also find many cases of mutual ornaments that cannot be explained by this model. Patterns of gains and losses of ornaments in either sex are highly idiosyncratic. The authors highlight the possibility that some (perhaps many) cases of mutual ornamentation may still be maladaptive, even if they are not the result of genetic correlation as originally envisaged. Physiological constraints could cause maladaptive ornament expression in females as a pleiotropic effect of for example selection on hormone levels. The authors argue that constraints should be given more attention in empirical studies of mutual ornamentation.

In: Animal Behavior: New Research
Editors: E. A. Weber, L. H. Krause

ISBN: 978-1-60456-782-3
© 2008 Nova Science Publishers, Inc.

Chapter 1

SEXUAL SELECTION, MATE CHOICE, AND PRIMATE PREFERENCES

*Jack A. F. Griffey** *and Anthony C. Little[†]*

Department of Psychology, University of Stirling, UK

1. INTRODUCTION

1.1. Defining Mate Choice and Preference

Mate choice and the preferences that many animals display when selecting potential mates are, as Kokko *et al.* (2003) explain, incredibly important evolutionary processes which impose selection upon the opposite sex and are accountable for a vast array of spectacular ornaments and characteristics that remain inexplicable via natural selection alone (Darwin 1871; Andersson 1994). Consequently, due to the evolutionary importance of mate choice and perhaps, as Bateson (1983) suggests, the renewed vitality of areas such as evolutionary and population biology, since the 1970s sexual selection and its implications for mate choice have experienced a rapid revival in interest. Based upon a number of major theoretical insights and empirical findings (Birkhead and Møller 1998; Eberhard 1996) there has been a growing interest in the mating preferences of animals to the point where mate choice and sexual selection have been two of the most active disciplines of scientific research within behavioural ecology and evolutionary biology (Gross 1994).

However, before we may review the current literature concerning data on primate mate choice it is first important to define a number of the key terms used in this area of study, provide an overview of the mechanisms via which sexual selection may occur, review a number of the most prominent models currently proposed for the evolution of preference and mate choice and discuss the various benefits which may be obtained for both females, males and offspring through this process.

---

[*] For correspondence. School of Psychology, University of Stirling, Stirling, FK9 4LA, UK. Email: j.a.griffey@stir.ac.uk
[†] Email: anthony.little@stir.ac.uk

As Soltis *et al.* (1999) suggest, in any study of sexual selection and mate choice it is particularly important to explain the distinction between the terms 'mate choice' and 'preference'. For example, Soltis *et al.* (1999) suggest that the use of the term 'preference' when utilized within contexts concerning mate choice, most commonly refers to internal motivation towards certain mates or the internal expression of a mating bias, which can only be measured experimentally. They also suggest the term 'mate choice' can be viewed as the subsequent expression of this preference within a particular field of constraints, which may ultimately act to inhibit or alter these preferences. Therefore under the definitions of Soltis *et al.* (1999), while both terms appear synonymous to one another it may in fact be more useful and accurate in the following review to consider preference (for a specific trait or number of traits in the opposite sex) as a mating bias or driving force that results in the expression of a particular behavioural outcome that we know as 'mate choice' and which is in itself part of a larger evolutionary process known as sexual selection.

Previous studies across a variety of species have identified a large number of preferences for various traits in the opposite sex that have been shown (both experimentally and via observation) to play a significant role in the process of mate choice (see sections 1.8, 1.9, 3.). First we examine the theoretical background underpinning preference and mate choice.

1.2. A Brief History of Sexual Selection and Mate Choice

Central to any study of mate choice is a detailed understanding of sexual selection, first discussed by Darwin (1871). While Darwin (1871) proposed that natural selection acted as a mechanism to explain the selective force that an environment imposed upon an organism, he also recognised the selective nature that differential reproduction may have within the evolutionary process.

In particular, Darwin sought to explain a major problem in his theory of evolution via natural selection, namely why males across many different species of animal often possessed elaborate, conspicuous traits (a point perhaps most famously exemplified by male peacocks which possess large, ornate tail feathers) that would obviously result in a reduction in survival (Burk 1982). The solution that he came to was that these traits had evolved due to the competitive advantage they conferred to their owners during competition for mates (or mating opportunities), a process which Darwin (1871) termed 'sexual selection'. As Andersson (1994) suggests, the main goal of sexual selection, and indeed of this brief introduction, is to explain the evolution of traits that confer an adaptive advantage in the competition for mates, the mechanisms by which they are favoured (in this instance 'mate choice') and briefly touch upon the occurrence and variation in some of these traits among a number of different organisms.

Ultimately, sexual selection centres on the assumption that variation exists between potential mates and that it is those individuals who possess traits that make it easier to secure a mate that will have greater mating success and in doing so produce more offspring that successfully reach adulthood themselves and reproduce in turn. The potential to pass on genes into subsequent generations is usually termed 'fitness'). As a consequence of this variation in mate quality competition over prospective mates occurs which is, as Andersson and Iwasa (1996) suggest, the unifying aspect of all forms of sexual selection.

Although a relatively simple idea, sexual selection (or simply competition over mates) may occur in varieties of forms (or via a number of different mechanisms) that have a number of important implications for many different organisms (see Andersson 1994; Andersson and Iwasa 1996, for a review). However it is perhaps mate choice that is the mechanism of sexual selection that has attracted the most interest within the scientific literature (Andersson and Iwasa 1996). The role that mate choice plays within sexual selection and the subsequent implications and consequences that this has on the behaviour, morphology and life history strategies that organisms of both sex employ will therefore be addressed below as a comprehensive theoretical understanding of the pressures that sexual selection and mate choice place upon organisms will allow us to better understand the evolution of many of the exhibited mate preferences that we will discuss later in this review.

1.3. Mechanisms of Sexual Selection: Intra-sexual and Inter-sexual Selection

As previously explained, while primarily sexual selection is concerned with competition between individuals over mating opportunities/prospective mates, sexual selection may in fact occur in one of two forms, either intra- or inter-specifically (see Moore 1990). Intra-sexual selection occurs when members of one sex (most commonly males) compete with one another for access to the other sex for mating opportunities, while inter-sexual selection (or 'female choice') occurs when females choose males to mate with based upon certain traits or characteristics which signal to the female that this male possesses a certain quality which will either help the female produce more or better quality offspring. Ultimately this makes males in possession of these traits potentially more attractive to females and therefore more likely to successfully gain a mating opportunity. As will become apparent it is this form of inter-sexual selection that dictates the mating preferences that we will cover.

Interestingly this brief discussion of the various forms of sexual selection that may arise highlights an important point for consideration that has particularly serious implications in our understanding of the mechanisms underlying mate choice and preference, namely why is it commonly males who must compete for mating opportunities and the females who may be 'choosy'?

1.4. Choosy Females and Competing Males

Mate choice, as a mechanism integral to sexual selection (and in particular inter-sexual forms of sexual selection) is ultimately defined by the act of one sex choosing to mate with the other on the basis of certain attributes or qualities. However as briefly mentioned above, in the vast majority of cases it is the female sex that 'chooses' their mate while males compete with one another for mating opportunities. The reasoning underlying this asymmetry between the sexes and the theory suggested to support this will be discussed below.

1.5. Asymmetry between the Sexes

Asymmetry between the sexes extends to a far more basic level than simple differences in mating tactics (and in fact it is due to the asymmetries at this more fundamental level that such differences in mating strategies are employed). The central issues surrounding the differential mating strategies of either sex involve the occurrence of anisogamy and the enormous impact that this has upon the potential lifetime reproductive success of individuals of each sex.

Males and females across all species exhibit a high degree of anisogamy (a difference in the size of their gametes) whereby females produce large, immobile macrogametes known as eggs which are rich in energy, and males produce many small and highly motile microgametes known as sperm. It is proposed that the evolution of anisogamy arose due to two basic selection pressures, namely for increasing zygote size therefore improving the chances of zygote survival and for increasing total gamete number (see Andersson 1994; Hoekstra 1987 for a review). Females therefore invest inherently more in an offspring prior to fertilisation than males (for mammal's, internal gestation and lactation further increase the additional cost placed upon females prior to birth) and it is this initial asymmetry in investment which inevitably leads to sexual conflict and the differences in mating strategies employed between the sexes of each species (see Andersson 1994; Bateson 1993; Kappeler and van Schaik 2004; for a review). Of course in species where gestation or parental care is the province of females, then the asymmetry in investment can extend far beyond the point of conception.

Competition for mates (or rather sexual selection) is generally more pronounced in males because, as Bateman (1948) proposed, the strength of sexual selection depends upon the relationship between mating success (e.g. the number of mates) and offspring production (e.g. fecundity), a relationship referred to as Bateman's Principle. Although across both sexes the mean lifetime reproductive success of each sex must be equal the *variance* in potential rate of reproduction for individuals within each sex may differ significantly (see e.g., Clutton-Brock 2007). For example males possess many small gametes which they invest very little energy into and therefore are potentially able to sustain a much higher fecundity than their female counterparts whose reproductive output is constrained by the production of much larger, more energetically expensive gametes and the responsibilities of postnatal child care (see Andersson and Iwasa 1996; Trivers 1972). Therefore as males, relative to females, provide a reduced investment (in terms of gamete production and parental care) they are less constrained in the number of individuals they may plausibly mate with, which ultimately increases their potential rate of reproduction (PRR) (see Clutton-Brock 2007; Clutton-Brock and Parker 1992; Ahnesjo *et al.* 2001) and biases the relative numbers of sexually active males to receptive females within a population. This leads to stronger selection pressure on males to acquire mating opportunities and thus increases the intensity of intra-sexual competition and the selection for sexually selected secondary traits in males rather than in females in order to attract potential mating partners (Emlen and Oring 1977). Females however are limited in the number of offspring they can produce due to the increased amount of pre- and postnatal investment they must provide (gamete production and postnatal care). Therefore in their case, selection favours the evolution of 'selectivity' in mate choice, which in turn generates additional selection pressure for male possession of secondary sexual traits that may signal their quality as mating partners to females (Trivers 1972). Ultimately then it is due to initial differences in gamete production and parental investment between the sexes that results in the asymmetrical sexual strategies that each sex employs as male and females are subject to different degrees of selection pressure (imposed by sexual selection) which act to influence each sex separately (both morphologically and behavioural).

The pressure placed upon males to acquire a large number of mating opportunities results in male-male competition for females and, as will be demonstrated, a vast array of behavioural (e.g., vocalisations) and morphological adaptations (e.g., tail length, markings, and colouration in Peacocks) designed to attract the attention of potential females mates and advertise the male individuals quality as a suitable mating partner. Females on the other hand, who may only mate with a limited number of individuals within their lifetime, can instead afford to be 'choosy' and show preference for mates who display various traits and characteristics that signal an individual's mate's potential quality (see Bateson 1983 for a detailed review).

1.6. The Evolutionary Importance and Implications of Mate Choice

As discussed, in situations regarding mate selection, females can be selective in their choice of mate and show preferences for the male possession of specific traits or characteristics that display to females some indication of the individual male's quality as a breeding partner (Trivers 1972). However for female preference to be considered adaptive the traits and attributes that males display to potential choosy females have to actually confer some form of direct evolutionary benefit to the selecting female.

While numerous correlation and experimental studies across many taxa have successfully demonstrated female preference for exaggerated and extravagant males traits (see Andersson 1982, 1994; Bakker and Pomiankowski 1995; Bradbury and Andersson, 1987; Ryan and Keddy-Hector, 1992; Møller, 1994a; Johnstone, 1995), it is, as Pomiankowski *et al.* (1991) suggests, of the utmost importance to first understand how (i.e. via what mechanism?) and why (i.e. what evolutionary benefits does this confer to the individuals involved?) these preferences for various exaggerated male traits may have developed and evolved if we are to fully understand and appreciate the evolution and importance of such studies with regards to mate preference and mate choice.

Therefore before we attempt to highlight specific examples of mate choice within the current scientific literature this review will address 3 hypotheses proposed as suitable selection pressures and mechanisms responsible for the evolution of female preference as discussed by Kirkpatrick and Ryan (1991) (see also Bulmer 1989; Jennions and Petrie 1997; for a review) and an additional hypothesis discussed by Barrett *et al.* (2002). We will then briefly describe a number of benefits conferred to females via mate choice.

1.6.1. Mechanisms Driving the Evolution of Mate Choice
Kirkpatrick and Ryan (1991) review two separate types of hypotheses for the mechanism driving the evolution of mating preferences, namely direct and indirect selection of preferences. Direct selection includes preferences which may confer immediate and direct benefits upon the selective female (i.e. increase female survival or fecundity) while indirect selection of preferences include a number of mechanisms proposed to select for certain female preferences including Fisher's (1930) runaway process of selection, the parasite

hypothesis (also know as the Hamilton-Zuk hypothesis; Hamilton and Zuk 1982) and finally Zahavi's handicap Principle (Zahavi 1975, 1977; Zahavi and Zahavi 1997)). As we will see both the Hamilton-Zuk hypothesis and Zahavi's handicap principle (sometimes both referred to as 'good genes' explanations of selection; see Jennions and Petrie 1997) differ significantly from Fisher's process of runaway selection though both may still provide indirect forms of benefit to 'choosy' females.

1.7. Direct Selection of Preference

Based upon a number of findings that have emerged from theoretical studies conducted in this area, direct selection of mating preference is thought to arise in situations where female preference is seen to affect survival or fecundity, and is specifically favoured as a form of selection in situations that directly increase the fitness of females displaying a certain preference (Kirkpatrick and Ryan 1991). As will be discussed later in more detail, direct selection is thought to occur via a variety of mechanisms including selection based on male resource provision (towards a female or offspring, Kirkpatrick 1985), advantages brought about by a reduction in costs incurred via searching for mates (Parker 1983; Anderson 1986; Pomiankowski 1987), selection against hybridisation (see Sanderson 1989; Lande 1982) and selection based upon differences in male sperm fertility (Grafen 1990a). Interestingly once a mating preference is established via direct selection it ultimately dictates the equilibrium for the evolution of the male trait to a point that although adaptive from a female perspective, female preference for a specific trait may in fact result in the evolution of male characteristics which may ultimately decrease male survival (see Kirkpatrick and Ryan 1991).

1.8. Indirect Selection of Preference

1.8.1. Fisher's (1930) Runaway Process of Selection

Runway selection was an idea first proposed by Fisher (1930) and, unlike the mechanism described above, is an indirect form of selection pressure that may influence the *total* fitness of the female exerting the preference (Kirkpatrick 1996). Fisher (1930) suggested that if both the male trait in question and the preference for this trait were both genetically determined then increased intensity in female preference for a specific trait would, in turn, lead to an increased exaggeration of the male trait in question and that this process could form a positive feedback loop, or a 'runaway' process, whereby over time female preference could greatly exaggerate a particular male characteristic or trait, potentially even to the maladaptive extreme whereby the trait evolves to a point where it impacts upon the survival enough to exactly balance the mating advantage that it confers to the individual (Andersson 1982, 1986; Pomiankowski 1987). Importantly this is a process of selection entirely dependent upon heritability as it requires offspring to successfully inherit either their parent's preference (daughters), or trait (sons), if the trait and preference is to exaggerate and propagate successfully over evolutionary time. This is an idea known as the 'sexy sons' hypothesis (Weatherhead and Robertson 1979), which suggests that individuals may indirectly benefit their own fitness simply by producing offspring (sexy sons) who will themselves will be highly successful in attracting mates if female preference for a particular male characteristic is similarly transmitted heritably to female offspring (see Kirkpatrick 1985; Pomiankowski *et al.* 1991). If so, these 'sexy sons' will go on to produce large numbers of offspring themselves, which in turn, indirectly benefits their parent's own fitness. It is for this reason that this type of selection is 'indirect' as a female (and in turn a male) may increase their inclusive fitness merely by mating with a male who possesses a trait that does nothing else but makes him (and therefore any resulting male offspring) attractive to females (see Pomiankowski *et al.* 1991).

1.8.2. The Genetic Heritability of Preference

As previously mentioned, female preference for elaborate male traits have been experimentally and observationally demonstrated, however as discussed above, most theoretical models of selection i.e. (e.g. Fisher, 1930; Lande, 1981; Iwasa *et al.* 1991) regarding the evolution of female choice also tend to assume a heritable basis to mating preference. Therefore, in order to validate these models of selection it is also of vital importance that studies are conducted which successfully demonstrate that the female preference for, and male acquisition of, a particular trait do indeed genetically co-vary and are heritable, as without this theoretical models such as Fisher's (1930) runaway process and 'good genes' explanations of sexual selection (see sections 1.8.1, 1.9) simply cannot act as forces maintaining female preference (Bakker and Pomiankowski 1995; Boake 1989; Bakker 1990; Ritchie 1992). Fortunately, a number of such studies have been successfully conducted (for a comprehensive review see Bakker and Pomiankowski 1995; Jennions and Petrie 1997).

For example in an classic experiment conducted by Bakker (1993) using Three-spined Sticklebacks (*Gasterosteus aculeatus*) it was demonstrated that both male colouration (the males of this species show conspicuous red colouration) and female preference for this, genetically co-varied i.e. that both female preference and the display of male sexual signals co-evolve with one another and importantly, that daughters preference for 'redness' and intensity of redness in sons, obtained via a breeding design, were also found to genetically co-vary: redder fathers were found to produce redder sons and a daughters preference for redness in males was found to correlate with that of their mothers preference. Importantly, this result neatly demonstrates not only the positive genetic correlation that exists, at least for Three-spined Stickleback (*Gasterosteus aculeatus*), between male secondary sexual characteristics and female preference, but also the heritability of this correlation in progeny too (see Figure 1.).

Similarly, several other studies have tested for heritable mating preferences typically by selecting for specific mating preferences across a variety of organisms (Jennions and Petrie 1997). These include early experimental manipulations by Majerus *et al.* (1982) into the female mating preferences of the Two-spotted Ladybird (*Adalia bipunctata*) who were able to demonstrate preferential mating in the female of this species and its role in the maintenance of colour polymorphism, whereby the population showed significant increases in the proportion of females mating with melanistic males over time, indicating a heritable basis to this mating preference within this species (note however that attempts to repeat the results of this initial study using both wild stock and laboratory based populations of Two-spotted Ladybird have failed (Kearns *et al.*, 1992)). Similar studies conducted into the genetic underpinnings of female preference have also demonstrated additive genetic variation in the mating preferences of Guppies (*Poecilia reticulate*; Houde 1994) Fruitflies (*Drosophila melanogaster*; Kaneshiro 1989; *Drosophila mojavensis*; Koepfer 1987); Grasshoppers

(*Chorthippus brunneus*; Charalambous *et al.* 1994) and Planthoppers (*Ribautodelphax imitans*; De Winter 1992).

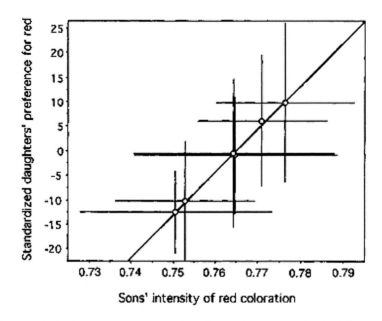

Figure 1. Positive correlation between ornament and preference in progeny obtained from a breeding design demonstrating both the heritability and covariance of trait and preference in Three-spined Stickleback (*Gasterosteus aculeatus*). (Taken from Bakker 1993).

Additional studies, similar in design to those of Bakker (1993) that utilise breeding designs and lineage analysis (parent-sibling, half-sib/full sib comparison) to ascertain a genetic basis to preference have also been conducted. For example, parent-sibling analyses conducted by Moore (1989) into the pheromone-based mating preferences of Cockroaches (*Nauphoetia cinerea*) indicate a genetic basis to preference in this species. Interestingly, lineage analyses conducted by Roelofs *et al.* (1986) on a particular species of moth (*Argyrotaenia velutinana*) indicate a similar, heritable, male mating preference for pheromones in potential female mating partners. However, other studies conducted into female preference find little or no such evidence for a genetic component in female preference. For example, mother-daughter analysis conducted by Johnson *et al.* (1993) into the mating preferences of Red Junglefowl (*Gallus gallus*) found no evidence of heritability in the mating preferences displayed by mothers and their offspring. Similarly Nicoletto (1995) reported no evidence of heritability in female preferences for male colouration in Guppies (*Poecilia reticulate*). Interestingly however, a number of studies have also experimentally applied artificial forms of selection pressure for certain traits and have recorded similar correlated changes in preference indicative of positive genetic correlation between preference and the male trait (see Jennions and Petrie 1997; Pomiankowski and Sheridan 1994; for a comprehensive review of these studies).

The current literature suggests then that both the evolution of female preference for and male possession of, a particular sexually selected trait do indeed co-vary with respect to one another and are, at least in some cases heritable. Fisher's (1930) runaway process of selection therefore remains the standard explanation for the evolution of exaggerated female preferences and male secondary sexual characteristics/traits (Pomianowski *et al.* 1991), however mathematical formulations of the runaway process find that it is unable to account for the stable exaggeration of female preference if this preference carries a cost with it (see Pomianowski *et al.* 1991; Pomianowski 1987; Bulmer 1989). Consequently other forms of selection pressure must exist that function to facilitate and maintain the evolution of seemingly 'costly' female mating preferences. Ultimately such theories must confer some form of indirect benefit upon the selecting female (see section 2.) in order to counteract the potentially detrimental effects of selectively choosing mates rather than simply maximising the potential mating opportunities (and therefore number of offspring) that an individual may have within their reproductive lifetime (see Kokko *et al.* 2003). Two such theories have been proposed which both incorporate Fisherian selection for runaway characteristics though in addition discuss potential mechanisms which may signal to females the genetic quality (i.e. "good genes") of their potential mate and thus may allow females to gain significantly more (in terms of total fitness) from their preferences for mates. Therefore the following two models may allow female mate choice to evolve even when significant costs are imposed upon the selecting females as a consequence of being 'choosy' (Andersson 1986; Pomianowksi 1987).

1.9.'Good Gene' Models of Selection

1.9.1. Zahavi's (1975) Handicap Principle

Briefly explained, Zahavi's (1975) handicap principle suggests that the ultimate benefit conferred via female mate choice and preference lies in the increased offspring survival it may facilitate via selection of potentially high quality mates (Pomiankowski *et al.* 1991). As Iwasa *et al.* (1991) explain, the handicap principle suggests that elaborate male ornamentation, a central feature within any form of sexual selection, in fact acts to provide information about the heritable genetic quality of the males themselves. In turn this allows females to mate preferentially with males who possess 'good' (and importantly heritable) genes that will indirectly benefit the survival of any resulting offspring.

Elaborate male traits may act as costly-to-fake or 'honest' signals of genetic quality to prospective female mates as the development and costs imposed via possession and maintenance of such elaborate traits (i.e. energetic, viability) require mates to be of a certain quality (Zahavi, 1975). A small ornament that does not result in a handicap to the individuals chance of survival may be produced by all males, irrespective of their quality however in order to produce a large ornamental trait, that has significant negative impact upon the survival of the individual (and additionally in terms of energetic demands), requires a viable and high quality organism (Iwasa *et al.* 1991). Therefore mate preference is favoured by selection if it is for male traits or ornaments that handicap the survival of the individual as only those males of true genetic quality can survive until maturity despite the costs imposed by the particular handicap (i.e. a long tail). Importantly a number of studies have found that these handicaps *must* be costly to produce and to maintain in order that such traits may remain as honest indicators as to a potential mates quality (Zahavi 1977; Grafen 1990b). Under the assumptions of Zahavi's (1975) handicap hypothesis as certain male traits may act as truly honest and costly-to-fake indicators of potential quality, females should actively show a preference towards those males with larger, more elaborate traits as these are indicative of males of higher genetic quality. Ultimately a female who chooses to mate with such an individual will indirectly benefit from such mate choice as these males will pass their heritable genetic quality ('good genes') onto offspring increasing their chances of survival in the future (Iwasa and Pomiankowski 1994; Zahavi and Zahavi 1997).

To date, a number of theoretical studies have established that Zahavi's (1975) handicap principle works (for reviews see Harvey and Bradbury 1991; Maynard Smith 1991). For example, Grafen (1990b) has shown that indicator mechanisms can favour the evolution of costly male ornamentation and female preference for these, in the absence of a Fisherian process. In addition, several experimental studies have also demonstrated mate choice based upon ornaments proposed to signal 'good genes' to potential female mates. For example, in an early study conducted by Maynard Smith (1956) it was found that female Fruit flies (*Drosophila subobscura*) often avoided mating with and rejected genetically unfit males (i.e. those that were highly inbred). Interestingly, these highly inbred males were unable to perform the normal courtship 'dance' and females who bred with outbred males (i.e. males with greater genetic quality) were found to produce many more viable offspring. This evidence not only implies a female preference for high quality mates on the basis of a particular elaborate male characteristic (courtship dance) but also demonstrates the indirect benefit that a female may gain from mating with a male of greater genetic quality (more viable offspring). In an experiment conducted by Norris (1993) on Great Tits (*Parus major*) it was found that females preferred to mate with males who possessed larger black breast strips. A series of cross-fostering experiments revealed that male strip size was heritable and that there was a strong positive relationship between the size of the father's stripe and the number of male offspring that survived within a brood (interestingly benefit in terms of survivorship were not passed onto female offspring). Other similar examples include experimentation by Møller (1994c) who successfully demonstrated a correlation between male ornamentation (specifically tail length) and resulting offspring viability in Barn Swallows (*Hirundo rustica*) and Petrie (1994), who found a significant interaction between male peacock (*Pavo cristatus*) attractiveness (assessed via the mean area of the fathers eye-spot on their train) and the size of offspring at 84 days and also survivorship of these offspring after 24 months. Studies such as these then appear to offer some of the strongest support in favour of Zahavi's (1975) handicap hypothesis suggesting that via the process of mate choice, and specifically preference for elaborate and costly male traits or characteristics, females obtain heritable viability benefits for their offspring (Krebs and Davies 1997).

1.9.2. Hamilton-Zuk Hypothesis (1982)

The handicap hypothesis, however, is not the only the model that incorporates the assumption of preferential mating for the indirect benefit of 'good genes'. Another theory proposed to explain the evolution of female mate choice and preference, known as Hamilton-Zuk or the 'parasite hypothesis' (Hamilton and Zuk 1982), also centres on the evolution of male secondary sexual traits, and female preference for them, as a function of the genetic advantages they advertise to females, and more specifically the indicative role they play in signalling genetic parasite resistance, another large class of heritable genes that may be particularly attractive to 'choosy' females.

Central to the Hamilton-Zuk (1982) hypothesis is the suggestion that the genetic cycle of resistance that exists between parasites and hosts act to maintain substantial heritability of fitness upon which sexual selection, and in particular female preference, can act. Ultimately this theory is based upon initial experimental findings from a comparative study conducted by Hamilton and Zuk (1982) into blood parasites and their effects on the brightness of plumage colouration (and song variety and complexity) in several North American passerines which suggests that bright plumage (and male song) in these birds acts to indicate genetic resistance to parasites (as Hamilton and Zuk (1982) found increases in parasite load led to a reduction in brightness of male plumage colouration and song variety and complexity). As females should show a preference for resistant mates (and in turn their resistance genes) as these males will increase their offspring's viability due to inherited resistance (Krebs and Davies 1997) plumage colouration therefore acts, in a manner similar to those elaborate male traits in Zahavi's handicap hypothesis (1975), as an honest, and ultimately costly-to-fake, signal of a potential mates genetic quality.

1.9.3. Evidence for the Hamilton-Zuk Hypothesis (1982)

Since the initial experimentation conducted by Hamilton and Zuk (1982) a number of comparative and single species studies have been conducted. For example in a study carried out into the mating preferences of the Ring-necked Pheasant (*Phasianus colchicus*) and designed to test the assumptions of the Hamilton-Zuk hypothesis, Hillgarth (1990) found that male resistance to disease and parasite load could indeed be heritable and that a significant correlation existed between male display rate, parasite load and the mate choice of females in this species. Similar findings have also been made by Zuk *et al.* (1990) in an experiment conducted upon captive flocks of Red Jungle Fowl *(Gallus gallus)*. In this instance, Zuk *et al.* (1990) experimentally infected Jungle Fowl with an intestinal nematode and measured the parasites adverse effects upon the male secondary sex characteristics and female preference. Zuk *et al.* (1990) found that infected chicks grew more slowly than uninfected controls (particularly their comb length, an ornamental secondary sex characteristic) and possessed shorter and paler tail feathers than the uninfected control group. Females appeared to prefer uninfected males over infected males in a ratio of 2:1 and analysis of covariance revealed that female hens were using the traits on which the two groups differed (i.e. length and quality of tail feathers and comb) to make their mate choice decisions. These results suggest that parasite infection has a disproportionately larger effect upon the quality of secondary sexual rather than non-ornamental characteristics and, in line with the assumptions of the Hamilton-Zuk hypothesis, that a parasites diminishing effect on these secondary sexual characteristics has a significant impact upon female mate choice in this species. Although a number of findings comparable to those of Zuk *et al.* (1990) have also been made by Clayton (1990) using parasitized Rock-Doves (*Columba livia*) and by Houde and Torio (1992) in the colouration and female choice of parasitized Guppies (*Poecilia reticulata*) (see also McMinn *1990;* Kennedy *et al.* 1987) it is a series of experiments conducted by Møller (1990) (see also Møller 1994b, 1994c) on Barn Swallows (*Hirundo rustica*) which first identified that parasite resistance was linked to both the exaggeration of male traits and the increased survivorship of offspring brought about by female choice for these less parasitized males. Firstly, in an earlier experiment conducted by Møller (1988) it was found that female barn swallows showed a mating preference for males with longer tails and that they possessed fewer parasites. Secondly, via a series of cross-fostering experiments, Møller (1990) was able to show that males with longer tails produced offspring with much lower parasite loads than males with short tails (Figure 2.) and it was also shown that the number of mites that a male parent possessed correlated with the subsequent parasite load of their offspring (Møller 1990).

Finally, by artificially manipulating parasite loads in certain male individuals, Møller (1990) was also able to show that parasite load had a detrimental effect on growth rate, and therefore the survival of offspring. Ultimately Møller's experiments provided support for three major assumptions of the Hamilton-Zuk hypothesis namely, that parasites directly affect the fitness of their hosts (e.g. tail length), that there is heritable variation in parasite resistance, and that expression of a particular sexual ornament varies with parasite burden. Importantly Møller (1990) also successfully demonstrated that females use this variation in expression of the male trait during mate choice in order to produce offspring with the greatest fitness potential possible brought about by heritable genes resulting in lowered parasite loads.

1.9.4. Fluctuating Asymmetry (FA) and Mate Choice

Finally, we briefly mention here an additional trait proposed in the assessment of mate quality, and in particular 'good genes', which unlike plumage brightness or quality (which is simply indicative of a single type of 'good gene' such as parasite resistance) is based upon a phenotypic measure which may indicate overall genetic quality (Krebs and Davies 1997). This measure is known as fluctuating asymmetry (FA) and typically refers to any deviation from perfect symmetry in bilaterally symmetrical traits (Van Valen 1962). As suggested by Soulé (1982) such departures from symmetry are assumed to be the result of environmental stressors (e.g. disease, parasitic infection), which destabilise those developmental processes encoded within our genes that lead to the development of symmetrical bodily traits and features. Therefore, a greater degree of symmetry in a particular trait (e.g. tail feathers) should signal to potential mates the presence of 'good' genes in an individual as it suggests that they possess genetic quality which is able to withstand and resist a number of environmental pressures resulting in the production of a symmetrical trait (Manning 1995; Watson and Thornhill 1994).

For example in an experiment conducted by Møller (1992) it was demonstrated that the mate preferences displayed by female Barn Swallows (*Hirundo rustica*) may in fact be correlated with male fluctuating asymmetry as parasite load is know to increase the level of FA in tail length and symmetry which ultimately influences the attractiveness of males as potential mates in this species. Similarly, in a study conducted by Thornhill (1992), it was found that the amount and/or quality of a pheromone produced by the Japanese Scorpion fly (*Panorpa japonica*) which is used to attract mates, is correlated with the fluctuating asymmetry of various male morphological traits (e.g. wing length) in this species. Furthermore, an experiment conducted by Møller and Höglund (1991) demonstrated that when compared to other morphological features, sexually selected characteristics and traits (in this instance, tail length in 16 bird species) tended to display higher levels of FA suggesting that signalling ones degree of FA is a particularly important function of sexually selected traits over other morphological features. Therefore this evidence, along with experimental findings in other species, is taken to suggest that fluctuating asymmetry in sexually selected traits is a reliable indicator of a potential mate's genetic quality (see Krebs and Davies 1997).

In summary, a number of mechanisms have been proposed for the evolution of female preference including direct and indirect selection, runaway and good genes hypotheses of selection. However it is important to remember that while these forces are often portrayed as mutually exclusive from one another this may not be entirely accurate. As Krebs and Davies

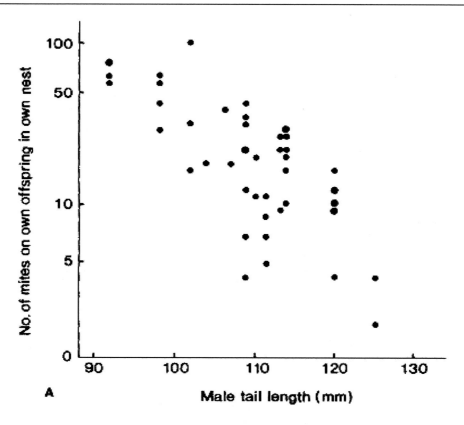

Figure 2. Negative correlation between male tail length and their subsequent offspring's parasite load in Barn swallows (*Hirundo rustica*) (taken from Møller 1990).

(1997) suggest, there are numerous scenarios in which various selective forces for female preference may interact with one another influencing preference evolution, the challenge now lies in understanding the relative importance of each of these selective forces in the mate preferences we observe and identifying how these different forces may interact with one another to influence mate choice and the evolution of female preference.

Finally, before we review a number of species-specific mate preferences it is important to highlight additional benefits, other than the selection of 'good' genes for ones offspring, which individuals may gain through the active selection of mates during mate choice.

2. THE BENEFITS CONFERRED VIA MATE CHOICE

In a detailed review Andersson (1994) considers a number of potential benefits conferred upon females, besides the advantage of simply mating with another individual and the acquisition of 'good genes' (see section 1.9), which may account for the costly practice of female discrimination of potential mates, otherwise known as female mate choice.

2.1. Mate Choice for Fecundity

Female mate choice and the benefits conferred from its practice may be based upon differences in the potential fecundity or fertility of prospective mating partners. For example as Williams (1992) suggests, if males vary in their fertilization ability (i.e.. differences in sperm supply) then females may indirectly maximise their fitness by mating with the most fertile of males thereby reducing the risk of producing infertile eggs and increasing their potential fecundity. Interestingly, a number of experimental studies appear to confirm this female preference for fertility. For example, a study by Robertson (1990) into the mating preferences of the Australian frog (*Uperolia laevigata*) suggest that females show a preference for males of a certain size which leads to high fertilization success. Similarly, experiments conducted on stocks of Lemon Tetra (*Hyphessobrycon pulchripinnis*) indicate that females show a preference for males with increased sperm supply as females appeared to prefer to mate with those males who had not spawned recently (Nakatsuru and Kramer 1982). Rate of male display has also been found to correlate with sperm supply in a number of species including Smooth newts (*Triturus vulgaris*; Halliday 1976) and Checkered White butterflies (*Pieris protodice*; Rutowski 1979). The natural occurrence of multiple matings by females across many species has also been suggested to occur due to the reproductive advantages it confers upon females practicing it. For example, Madsen *et al.* (1992) demonstrated in a small, inbred, group of Adders (*Vipera berus*) that the proportion of viable offspring produced by a female was proportional to the number of mates that the female engaged with. Similarly, Olsson (1992) demonstrated in a population of Sand lizards (*Lacerta agilis*) that the proportion of viable eggs produced was dependant upon the number of matings that a female had. Both studies therefore highlight the important role that fecundity and fertilization ability may have in the process of mate choice.

Mate choice for fecundity in females may also be particularly advantageous during male choice of mate (particularly in species where females differ markedly in size, a characteristic thought to be particularly indicative as to a female's fecundity; see Andersson (1994) for a review). This may arise because mating incurs large costs on the male as well as the female in terms of energy, time, sperm depletion, and a reduction in the potential to fertilise other females (Andersson 1994), therefore males should show a preference towards the most fecund females as mating partners in order to indirectly increasing their own fitness by maximising their potential for producing offspring (Parker 1970). Consequently, as may be expected, males show a strong preference for fecund females during mate choice. For example, a study conducted by Gwynne (1981), into the Mormon cricket (*Anabrus simplex*), a species where the female mounts the male prior to copulation, identified that in approximately two-thirds of the 45 cases of pre-copulatory mounting observed, the male pulled away from the female prior to the transfer of the male spermatophore. In this instance it is suggested that males are able to assess the mass of mounting females from which they may infer the fecundity of the female with whom they are mating with which, as Gwynne (1981) suggests, confers upon selective males a fecundity advantage of approximately 50%.

2.2. Mate Choice for Nutritional Benefits

As Andersson (1994) suggests, in addition to the benefits acquired in terms of fertility and fecundity, female preference for certain males may be advantageous in terms of nutritional benefits that males may offer to females. Such nutritional benefits may appear in a variety of different forms including prey, seminal nutrients or even to the extreme during suicidal food transfer where the male offers themselves up to the female be eaten (an act perhaps most famously demonstrated by the Praying Mantis (*Mantis religiosa*; see Roeder 1935).

Interestingly, a number of studies conducted into bird and insect courtship or nuptial feeding (the gathering and offering of food to mates by males) have found that this practice can act to enhance female fecundity (Thornhill 1983; Carlson 1989; Simmons 1990; see Andersson 1994 for a review). For example, in a review conducted by Thornhill and Alcock (1983) it was demonstrated that a number of insect females choose a mate based upon his courtship feeding and that consequently these females reproduce better as a result. Similarly, a study conducted by Tasker and Mills (1981), found that for the Red-Billed gull (*Larus novaehollandie scopulinus*), the likelihood of copulation after courtship increases if the male feeds the female. Interestingly, it has also been found that the rate of male courtship feeding in the Common Stern (*Sterna hirundo*) correlates with later rates of feeding the young (Wiggins and Morris 1986), suggesting that females may use feeding behaviour as a reliable indicator to the males parenting quality also. Nutritional benefits acquired via mate choice may also be obtain via seminal fluids which provide females with an extra source of nutrition prior to development of the egg (Markow 1988; for review see Andersson 1994). Ultimately this may benefit the fecundity of the female (Butlin *et al.* 1987) and it is thought that the transfer of nutrients such as these which are synthesised by the males may in part represent a mating effort that raises the males chances of fertilising eggs (Andersson 1994).

Interestingly, many of the examples previously discussed demonstrate mate choice based upon assessments of a males personal attributes or genetic quality (i.e. 'good genes' hypotheses; fertilization and fecundity benefits; male nutrient synthesis) however mate choice may also be made based upon a number of non-genetic resources that a particular male has to offer as these are often factors which have significant implications in the chances of survival of any subsequent offspring produced from a mating opportunity. Therefore two of the most prominent forms of non-genetic benefit conferred to females via mate choice will be briefly discussed.

2.3. Mate Choice Based upon Parental Ability

As males often differ in their parental ability across many species, females may indirectly benefit via increasing the likelihood of their offspring's survival by choosing to mate with males who possess a greater parenting ability. For example, in a study conducted by Brown (1981) it was shown that female Mottled Sculpins (*Cottus bairdi*) showed a mating preference towards larger males. This is because larger males are known to be better at guarding and defending the nest than smaller males and therefore increase the offspring's chances of survival. Petrie (1983) observed that female Moorhens (*Gallinula chloropus*) demonstrated a preference for larger, fatter male mates who possess greater energy reserves than smaller, thinner males and who incubate more than others thereby enabling females to produce more clutches per season (Andersson 1994). Similarly, Muldal *et al.* (1986) demonstrated in the Red-winged Blackbird (*Agelaius phoeniceus*) that male parental ability influences both the number and weight of fledglings produced.

2.4. Mate Choice Based upon Territory and Defended Resources

Female mate choice may also be based upon the potential benefits that a male's territory or possession of resources may offer to the female and to any offspring she may produce. For example Severinghaus *et al.* (1981) observed that in a certain species of bee (*Anthidium manicatum*) males defend flowers used for food and only permit females to feed from them if they mate with the male. In this species the amount of flowers that a male is able to defend correlates with the amount of females that the male subsequently attracts. Some species of fish also show a relationship between male mating success and territory. For example, Jones (1981) identified that female Wrasse (*Pseudolabrus celidotus*) prefer to mate with those males who possess territories in deep water. It is suggested that this is because this type of territory receives reduced levels of egg predation and therefore should increase the chances of offspring survival. Many species of birds also exhibit a strong relationship between male mating success and territory size or quality (for review see Andersson 1994). For example, Holm (1973) observed that those male Red-winged Blackbirds (*Agelaius phoeniceus*) that possessed territories that contained the most suitable or high quality vegetation for nesting also attracted the most female mates. While Collias and Collias (1984) found that female Village Weavers (*Ploceus cucullatus*) preferentially choose a mating partner based upon the quality of the nest that he builds. The relationship between territory quality and mating preference is also found in larger mammals too, for example, Kitchen (1974) found that in the Pronghorn Antelope (*Antilocapra americana*) males who have the best foraging opportunities within their territories attract and mate with more females than others.

Hopefully this brief review has demonstrated that mate choice may in fact lead to a number of benefits for the female other than simply the mating opportunity itself. These may be direct genetic benefits (via preference for mates that possess traits signalling genetic quality) or indirect benefits (i.e. nutritional, fecundity/fertility and resource/territorial) however both forms of benefit function in the same manner, to increase the likelihood of survival for the choosy female herself or the survival of offspring produced from a mating opportunity.

Up until now we have discussed a number of the mechanisms proposed to underlie the evolution of female preference and the benefits that may be conferred upon females via preferential mate choice. While many of the experimental and observation examples used so far to illustrate these points have covered a variety of species the remaining part of this review will now focus solely upon mate preferences displayed within the primate lineage. However, it is important to note that due to the vast amount of experimental and observational evidence regarding human mate choice it is more practical within the scope of the current review to focus solely upon the mate preferences displayed by non-human primates (for an extensive review of human mate choice see Geary *et al.* 2004).

3. NON-HUMAN PRIMATE MATE CHOICE

Despite the extraordinary colouration of the facial features of the male Mandrill (*Mandrillus sphinx*), one of the most sexually dimorphic primate species that exists today (Paul 2001), until recently, as Dixson (1998) has pointed out, traditionally when studying sexual selection and mate choice, primatologists have tended to concentrate solely upon intra-sexual competition (male-male competition over females) and the role that dominance and hierarchy plays when determining mating success within the mating systems of primates. However, over time, evidence has accumulated and recent theoretical insights and empirical findings into the study of mate choice (Birkhead and Møller 1998; Eberhard 1996) now suggest that in fact females across many primate species actively solicit copulations and that they show a preference for the 'type' of male they mate with, refusing to copulate with certain males while actively accepting others. The growing number of experimental and observational studies reviewed here has led to a considerable change in the way primatologists view the roles of males and females with regards to mate choice as evidence seems to suggests that females across many species of primate play a far more active role in the process of sexual selection and mate choice than previously thought and furthermore, that these mate choice decisions provide primates with a number of important direct or indirect benefits (Paul 2001). In fact as Paul (2001) and Birkhead (2000) suggest, with every new experimental findings it seems that the stereotypical view of aggressive, dominant males competing for access to passive females is increasingly being shown to be inappropriate. For this reason the remainder of this review will concentrate solely on the evidence for female mate choice as experimental evidence seems to suggest that within primate social systems it is predominantly females who exert mate preference and carefully choose their potential mating partners from a group of males competing for female access (Kappeler and van Schaik 2004) (although a number of cases of male mate choice have been documented, see Kappeler and van Schaik (2004) for review).

The issues regarding the deficiencies in our understanding of female mate choice in primates lie partly in the difficulties of measuring such preferences in a natural setting (Paul 2001), which ultimately has led primatologists to focus more attention towards the effects of male competition for female access on mating success (a behavioural variable far easier to observe) and has resulted in a situation where it now appears for example, that we know less about female choice in non-human primates than we do about the mating preferences of the Guppy or Widow bird (Møller 2000). Keddy-Hector (1992) concluded in a review of primate mate choice that the evidence for female mate choice in most species of non-human primates is modest at best, while others, such as Cords (1987) and Kappeler and van Schaik (2004), suggest that while female choice may be a particularly important force in the evolution of primate societies, at present there is surprisingly little, or only anecdotal evidence, for its existence. There are a number of reasons for these suggestions, for example, until recently (see section 3.2) there was a lack of experiments conducted with primates designed to test the predictions derived from sexual selection, very few studies had been conducted which incorporated both genetic paternity testing and behavioural analysis of sexual selection and mate choice, and finally, male and female reproductive interests, particularly in primates, are rarely similar and so it is extremely difficult to solely assess the relative importance of female preference in the process of mate choice (see Paul 2001). Fortunately however, and despite the numerous difficulties in conducting research into the preferences exerted by females during mate choice, when such investigations are successfully conducted the insight and understanding we may draw from the findings obtained are extremely interesting. Furthermore, as Kappeler and van Schaik (2004) explain, the potential importance of studies conducted into primate mate preference, and in particular female mate choice, are perhaps accentuated by the fact that primates have relatively slow life histories (females may then only produce a few offspring during their reproductive career, Ross 1998) and because the males of most primate species provide little or no direct parental care for offspring, consequently a female's choice in mating partner may have extremely significant effects upon her own fitness.

3.1. Evidence for Female Mate Choice in Non-human Primates

As previously noted, numerous primatologists (Dixon 1998; Keddy-Hector 1992; Paul 2001) now propose that females not only play a far more active role in sexual interactions but also display clear preferences for their mating partners, which suggests that female choice may be a far more powerful selective force among non-human primates than previously thought (Manson 1992). Despite the difficulty in assessing female primate mate preference a number of observational and experimental studies have been conducted and there is in fact, as Paul (2001) suggests, 'considerable evidence for both direct and indirect forms of mate choice in non-human primates.

One of the most commonly cited examples of direct female choice in primates is a preference for dominant males (Small 1989). For example, a study by Welker *et al.* (1990) into the mating preferences of the Brown Capuchin (*Cebus appella*) found that females from both captive and free-ranging groups showed a significant mating preference for copulation with the highest-ranking, most dominant male. A similar observation made by Janson (1984) describes the elaborate female courtships of wild Brown Capuchins (*Cebus apella*) whereby an estrous female will actively shadow a dominant male for days in order to solicit copulations from them using a series of vocalisations, hind-quarter presentations and grimaces. In the first laboratory experiments of female mate choice conducted on Vervet monkeys (*Cercopithecus aethiops*) by Keddy (1986), estrous females were placed within dyads with either high or low-ranking males and their subsequent behaviour was compared to their normal behaviour under group conditions. Interestingly, Keddy (1986) observed that all females preferred high-ranking, versus low-ranking males. In a study into the mating preferences of female free-ranging Vervets, Andelman (1987) also reported a similar preference for dominance whereby high-ranking males were found to possess a significantly greater amount of successful copulations than low-ranking males as females frequently rejected the copulation attempts of less dominant individuals. A number of studies have also demonstrated a female preference across many species of primate for males of physical superiority as signalled via various morphological or behavioural traits. For example, a study conducted by Boinski (1987) into the mating preferences of female Squirrel monkeys (*Saimiri oerstedi*) found females exhibited a significant mate preference for larger males (the largest male in the group studied obtaining 70% of the total copulations in one breeding season). Boinski (1987) suggested that such a preference for larger males might have arisen due to the Fisherian process of runaway selection or simply on account of the significant contribution that larger males make in terms of increased levels of vigilance and direct intervention in situations of possible threat relative to smaller males. Therefore Boinski (1987) argued that females chose males in anticipation of future parental care. Similarily, experimental observations of the mating preferences of female Orangutans (*Pongo pygmaeus*) conducted by van Schaik and van Hoof (1996) found that females display a mating preference for fully adult males with large cheek flanges and throat sacs. In a study conducted into female mate choice in a captive group of Japanese macaques (*Macaca fuscata*) Soltis *et al.* (1999) found that females during peri-ovulatory periods (when most likely to conceive), preferred males who displayed most frequently and interestingly, as Modhal and Eaton (1977) have shown that frequency of male display and of ejaculation in Japanese macaques to be significantly correlated to one another. This form of female preference then may have significant fitness benefits to the selective female.

Interestingly, however, a number of studies do not report a female preference for dominance or physical superiority. For example, in a study conducted by Smith (1994) and based upon long-term paternity data collected from a captive group of Rhesus macaques (*Macaca mulatta*), it was speculated that in fact females do not exhibit a preference for top-ranking dominant males but rather young males who may ultimately achieve a high ranking position within the social group. Studies conducted into Japanese macaques (*Macaca fuscata*) by Perloe (1992) also identified instances where females refused to copulate with high-ranking males in their troop, instead displaying a mating preference for younger males of a lower rank, a behaviour which Perloe (1992) speculates functions as an incest avoidance mechanism (e.g. avoidance of courtship with older males who may have previously mated with their own mothers). Similarly, Paul (2001) reports a number of systematic analyses of female proximity maintenance behaviour in Rhesus (*Macaca mulatta*) and Japanese macaques (*Macaca fuscata*) which suggests that in fact males attractiveness is not correlated with their dominance rank (see Manson 1992; Soltis *et al.* 1997). Therefore, as this brief review of the experimental and observational findings clearly demonstrates, currently the data regarding evidence of female primate preferences for male dominance is mixed, though it is interesting to note that it is macaque species, rather than other species of non-human primates, which appear to have generated somewhat different findings regarding their mate preferences perhaps indicating that there are potentially important species differences to explore for this species.

Female mate choice may also be based upon a number of male characteristics other than dominance. For example, in an idea first developed in studies of Olive baboons (*Papio anubis*) by Smuts (1985), it was suggested that female mate preferences could be based upon former affiliative relationships with males formed in non-sexual contexts, or phrased more simply, with 'friends'. Smuts (1985) found that certain female baboons formed long-term associations with particular males, grooming, sitting and even sleeping alongside them far more frequently than other males within the group. Interestingly, Smuts (1985) observed that these male 'friends' also experienced a greater advantage in mating opportunities with these females than did other males. It is important to note however that a number of more recent studies either have been unable to confirm this result (Bercovitch 1991) or indicate that these 'friendships' are typically established as a result of prior sexual activity (Palombit *et al.* 1997). Additionally, in sheer contrast to Smuts' (1985) 'friendship hypothesis', several other studies have also demonstrated a female mating preference for unfamiliar or novel males (Small 1989; Bercovitch 1997; but see Manson 1995).

Similarly, other studies have also revealed that females may simply show favouritism towards a particular male, which has significant implications upon the subsequent mate preferences they make. However, as Dixson (1998) suggests, what is particularly interesting is that often laboratory or observational studies are unable to explain the basis of this favouritism. For example, in a study of captive Chimpanzees (*Pan troglodytes*) by Yerkes (1939) it was noted that females tended to exhibit a preference for one of three males during pair tests to observe sexual interaction, indicating perhaps that females of this species may show favouritism or preference towards a particular male for no apparent beneficial reason. Similarly in a series of operant conditioning studies conducted by Eaton (1973) females were found to press bars in order to release certain males into their cage far more frequently than they did for other males. A number of similar studies demonstrating such favouritism during male mate choice have also been conducted (see Dixson 1998 for review).

As previously mentioned, the degree of parental care that a female receives from male partners across primate species is extremely low or absent (Kappeler and van Schaik 2004). Therefore it may be expected that, if possible to ascertain from behavioural information (i.e. male care towards a females offspring), females should show a strong preference towards males who care for their infants as this increased parental care may greatly influence a female's immediate reproductive success (Keddy-Hector 1992). In an experiment conducted on female Vervet monkeys (*Cercopithecus aethiops*) Keddy-Hector *et al.* (1989) examined whether females did in fact display a preference towards males who displayed parental care. In their experimental setup the behaviour of males were monitored in one of three isolated conditions (Plexiglass condition: female and male could observe each other; One-way mirror condition: only the female could observe the male; Metal partition: no observation between female and male). After a 30-minute observation period the female and male were brought together and their behaviour was monitored and recorded. Interestingly, Keddy-Hector *et al.* (1989) found that a female's behaviour towards a male was dependent upon the male's tenure, rank, and importantly, observed behaviour towards her infant. Female response was greatest towards subordinate males, with whom she displayed lowered levels of aggression as male affiliative behaviour towards her infant increased. Similar findings have also been made by Raleigh *et al.* (1992) who replicated the initial experiments of Keddy-Hector *et al.* (1989), however in this instance male care was directed towards another female's infant. Interestingly, Raleigh *et al.* (1992) also found that females still directed significantly higher affiliative behaviour towards males who cared for other female's infants. Together these experimental results suggest that females possess a significant preference for males who display parental care whether it is directed towards their own or another's infant as it potentially advertises favourable parental traits in these individuals (Keddy-Hector 1992).

Despite the fact that considerable variation may exist between individual primates mate preferences within a species (Paul 2001) one particularly important component of female mate choice appears to emerge throughout all major taxa of primate studied (see Paul 2001) due to its associated costs in terms of deleterious genetic effects upon offspring (Ralls and Balou 1982), namely a strong aversion to mate with close childhood associates or other family members, a behaviour commonly referred to as incest avoidance. Consequently as Dixon (1998) explains, due to the importance of this process for the genetic health of future offspring a great deal of observational and experimental evidence exists reporting that close relatives within primate social group actively avoid mating with one another (e.g. Goodall 1986; Kuester *et al.* 1994; Missakian 1973; Enomoto 1974; Packer 1979).

3.2. Experimental Manipulations of Primate Mate Preferences

Finally, we will consider in detail two studies recently conducted into primate mate choice and preference which more closely mirror a number of the experimental studies conducted in humans (*Homo sapiens*). Generally such studies aim to investigate the various preferences that we display during mate choice and involve experimental manipulations of particular traits and characteristics and investigation into the influence that this has upon the subsequent mate preferences we make. In this sense studies of this design are perhaps the most direct tests of many of the models of sexual selection and mate choice covered within this review (i.e. 'good genes' hypothesises) as they involve the identification and direct manipulation of certain characteristics proposed to have evolved due to sexual selection and subsequently identify that certain experimental manipulations of these traits result in a corresponding difference in the degree of preference shown towards them. Because of the unique design of these studies within the primate literature and the homology they display with respect to studies conducted into human mate choice they may particularly interesting and enlightening avenues for future studies of primate mate choice and therefore will be discussed in greater detail below.

3.2.1. Female Preference for Male Facial Colouration

Many species of primate possess conspicuous secondary colouration, for example, during the mating season both sexes of adult Rhesus macaques undergo, in particular, an increased reddening of the facial and anogential skin (Baulu 1976). However, the functional significance of much of primate secondary sexual colouration remains unknown. A number of studies have reported associations between intensity of colour and status within the group (e.g. Dunbar 1984; Gartlan and Brain 1968) and Waitt *et al.* (2003) suggest that this colouration might be associated with selection pressures imposed via female choice.

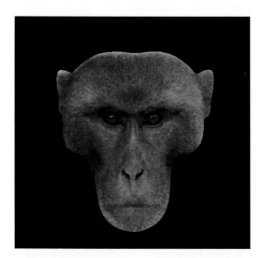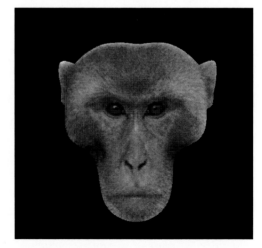

Figure 3. Same face colour transformations of red (left) and pale (right) versions of stimuli used by Waitt *et al.* (2003).

In an experimental study of Rhesus macaques (*Macaca mulatta*) conducted by Waitt *et al.* (2003) it was successfully demonstrated, via computer manipulation of red colour, that females exhibited a significant preference for red colouration in the faces of male conspecifics (see Figure 3.) whereby, overall, females spent significantly more time viewing red faces than pale faces. Amongst male rhesus macaques reddening of skin is regulated via testosterone, which is reported to have immunosuppressive effects (Folstad and Karter 1992), therefore Waitt *et al.* (2003) suggest that only those males in good condition may be able to endure the costs imposed via these colourful displays. Waitt *et al.* (2003) suggest that the female preference for red colouration observed in rhesus macaques may have arisen as preferential mating with these males, who possess such highly developed and costly displays, may be particularly beneficial to the female either via direct benefits to the female themselves through a reduction in pathogen transmission from potentially infected males (Loehle 1997) or indirectly beneficial by providing offspring with a heritable resistance to pathogens (Folstad and Karter 1992). As female Rhesus macaques are known to exhibit a high degree of mate choice that appears not to be based upon dominance rank (Manson 1994a) or upon affiliative relationships (Manson 1994b), the results of Waitt *et al.* (2003), which suggest female mate choice via physical features, and in particular colouration and the its associated advertisement of 'good genes', seem particularly plausible. Furthermore Waitt *et al.* (2003) suggest that the benefits conferred via this form of mate selection, namely pathogen resistance, may be particularly beneficial and relevant for Rhesus macaques as they possess a highly promiscuous mating system which ultimately leads to higher rates of sexually transmitted disease (STD) infection (Nunn *et al.* 2000). Waitt *et al.*'s (2003) study clearly demonstrates perhaps the first experimental evidence of the influence that male colouration may have upon male attractiveness to females and the subsequent effects that it has within a females mate choice decisions. It also provides particularly good evidence in support of a 'good gene' mechanism of selection underlying the mate choice and preferences of female Rhesus macaques.

3.2.2. Female Preference for Male Facial Symmetry

In a similar study conducted on adult Rhesus macaques (*Macaca mulatta*) by Waitt and Little (2006), the influence that bilateral symmetry of facial shape in assessments of attractiveness and mate choice were investigated. In humans a number of studies have demonstrated that facial symmetry may influence subsequent judgements of attractiveness in both real (Mealey *et al.*, 1999) and manipulated (Little *et al.*, 2001) faces, however, it is unclear if this is an effect unique to humans or whether manipulations of facial symmetry may also influence attractiveness and preference amongst other primate species.

It is suggested that the preference for symmetry may have arisen for a number of reasons including its role as an honest indicator of genetic quality and health (see van Valen 1962, section 1.9.4) or alternatively, due to perceptual biases in biological recognition systems (Enquist and Arak, 1994; Enquist and Johnstone, 1997) which suggests that the preferences for symmetry observed across many species have evolved not because of the underlying genetic benefits that such symmetry signals but rather due to a perceptual bias in the recognition systems of organisms which gives rise to a preference for symmetrical traits. Irrespective of the selective forces responsible for preferences for symmetrical facial stimuli very little work had been conducted into the evolutionary history of this preference and in particular no assessments have been conducted into the influence that facial symmetry had on the attractiveness of conspecifics, and the subsequent implications that this may have for mate choice decisions, in primates.

Waitt and Little (2006) conducted an experiment similar in design and methodology to that of Waitt *et al.* (2003) to assess whether manipulations of Rhesus macaque (*Macaca mulatta*) facial symmetry influenced subsequent visual preferences for opposite-sexed faces in this species. To experimentally assess this, Waitt and Little (2006) presented 13 adult Rhesus macaques with a number of computer manipulated images of symmetrical and asymmetrical versions of opposite-sexed conspecific faces (see Figure 4.). Waitt and Little (2006) employed looking behaviour (gaze duration and frequency) to assess 'visual preference' and found that overall, subjects looked significantly longer and more frequently at symmetrical rather than asymmetrical versions of faces indicating that Rhesus macaques, like humans, also show a preference for facial symmetry in conspecifics. This result not only implicates the potential importance of facial symmetry in assessments of potential mates but also, as Waitt and Little (2006) introduces "the possibility that human facial symmetry preferences are more deeply rooted in our evolutionary history than previously realized" (pp.139-140). Although this preference for symmetry may not necessarily determine all mate choice decisions, interestingly manipulations of symmetry did not appear to be equally influential between both sexes of macaque. Despite a non-significant interaction between sex and stimuli type in the analyses of Waitt and Little (2006), symmetry appeared to have a more important and substantial impact upon the preferences of female rather than male individuals. As a number of studies have highlighted the importance of female choice in primates (see section 3.1) the increased female preference for facial symmetry in male Rhesus macaques appears to be consistent with similar research into the mate choice decisions of other non-human primates.

 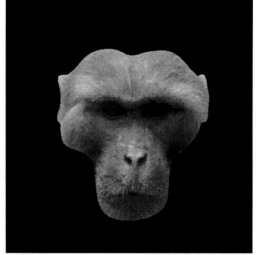

Figure 4. Example of symmetrical (left) and asymmetrical (right) version of female macaque facial stimuli constructed by Waitt and Little (2006).

In conclusion then, the study of Waitt and Little (2006) has particularly important implications, for future studies in the use of facial shape in assessments of primate mate choice decisions and preferences. In conjunction with findings from Waitt *et al.* (2003) it also seems that a variety of facial features may in fact influence the mating preferences of non-human primates in ways homologous to those previously observed in studies of humans mate choice.

Future research is needed to identify and understand the full range of traits and characteristics that interact to influence the preferences of non-human primates during mate choice, however, studies such as those conducted here appear to be a particularly successful and suitable means in which to do so. Further work will also need to be carried out in order to assess whether the preferences currently identified (e.g. symmetry; colouration) actually translate into preferences observed during real mate choice decisions. However, the current simplicity of these studies, the ability to accurately manipulate certain traits and characteristic and the direct and measurable effects that may be gained from their use indicate that studies of this design, homologous to a number of those conducted into human mate choice, may be a vital step in our future understanding of non-human primate mate choice.

5. SUMMARY AND CONCLUSIONS

The aims of this review were to address the current literature regarding sexual selection and mate choice and also to bring to light some of the most recent experimental and observational evidence for primate preferences and mate choice. Although primatologists have long been aware that non-human primates displayed particular preferences for certain mating partners during mate choice (Paul 2001) it is only recently that these preferences, particularly in females, have been addressed. While traditionally primatological examples of mate choice and preference relied solely upon observational evidenced and the inferences that one may draw from these regarding primate preference, a number of more recent studies (see section 3.2) have begun to experimentally assess primate mate choice via manipulation of certain stimuli thought to influence the mate choice decisions of prospective partners. In this sense these experiments more closely follow a number of mate choice experiments successfully conducted upon humans and indicate a potentially new and exciting means of understanding the mate choice decisions made by non-human primates. Not only are these types of experiments controlled, measuring the direct effects of manipulations of certain traits on primate preference, but interestingly, as they are also homologous to a number of studies conducted in humans, potentially they may allow us to more closely investigate the evolutionary history of a number of the mate choice decisions that we as humans display. Furthermore, more recent studies conducted into primate preference appear to focus their attention upon the function and evolutionary consequences of mate choice, which may also lead to interesting findings, particularly in relation to our understanding of primate sociality. For example, despite our initial assumptions, it seems that females do not simply mate with individuals based solely upon dominance hierarchies but in fact, across a wide range of species, base their mate choice decisions upon a number of factors which may have significant effects for their own survival and the survival of their offspring (see section 3.1).

As sexual selection is a powerful force acting on any species it is of vital importance that we fully understand the selective pressures and mate choice decisions that are being made if we are to fully understand and appreciate the full range of behaviours displayed by a particular species. Primates in particular (non-human and human), possess a form of female mate choice which appears to be a multidimensional process, requiring consideration and assessment of a number of male characteristics and factors including cultural, behavioural, and physical displays, which all aim to signal to females the male possession of a particular desirable trait. Consequently, it is only when we truly understand the significance and implications of all of these factors and their influence upon an individuals mate choice decisions that we will fully appreciate the behavioural manifestations of a particular species and the significant effect that a selective force such as sexual selection has upon the social and behavioural dynamics within a social group.

As is the case with many fields of study, the key in maximising our understanding of primate mate choice lies in the successful integration of data from the large number of observational studies conducted into primate mate choice along with findings gained from new methods of experimental study reviewed here. In doing so we hope to gain a broader and more thorough understanding of a particularly interesting and important area of primate behaviour, however, numerous questions remain to be answered. For example, what effect, if any, do experimental manipulations of facial masculinity have on primate preference? Do they mirror the effects found for humans? And similarly, what effect does manipulation of colouration or symmetry have upon the preferences of species of primate other than those investigated by Waitt *et al.* (2006) or Waitt and Little (2003)? Are the effects they obtained simply species-specific or can they be applied more generally across a number of primate species?

Potentially, as Keddy-Hector (1992) suggests, it may be that experimental manipulations are the best way to untangle such complex behavioural questions, however, as mentioned earlier, it is likely that the most accurate answers to such questions will come via integration of pre-existing observational data and through the careful design and implementation of experimental studies homologous to those conducted by Waitt *et al.* (2006) and Waitt and Little (2003). Based upon the rich variety of experimental and observational evidence reviewed here it appears that future investigation and experimentation into primate mate choice will be a particularly fertile and promising area of behavioural research. Via integration of observational and experimental evidence it is hoped that future work conducted in this area across a range of primate species will significantly increase our understanding and appreciation of a number of primate's sexual and social behaviours.

ACKNOWLEDGMENTS

Jack Griffey is supported by a PhD studentship from the University of Stirling. Anthony Little is supported by a Royal Society University Research Fellowship.

REFERENCES

Ahnesjo, I., Kvarnemo, C. and Merilaita, S. (2001). Using potential reproductive rates to predict mating competition among individuals qualified to mate. *Behavioural Ecology, 4*, 397-401.

Andelman, S. J. (1987). Evolution of concealed ovulation in vervet monkeys. *Amer. Natur.*, *129*, 785-799.

Andersson, M. (1982). Female choice selects for extreme tail length in a widow bird. *Nature, 299*, 818-820.

Andersson, M. (1986). Evolution of Condition-Dependent Sex Ornaments and Mating Preferences: Sexual Selection Based on Viability Differences. *Evolution, 40 (4)*, 804-816.

Andersson, M. (1994). *Sexual selection*. Princeton University Press.

Andersson, M. and Iwasa, Y. (1996). Sexual Selection. *TREE, 2,* 53-58.

Bakker, T. C. M. (1990). Genetic variation in female mating preference. *Netherlands Journal of Zoology, 40*, 617-642.

Bakker, T. C. M. (1993). Positive genetic correlation between female preference and preferred male ornamentation in sticklebacks. *Nature, 363*, 255-257.

Bakker, T. C. M. and Pomiankowski, A. (1995). The genetic basis of female mate preference. *J. Evol. Biol., 8*, 129-171.

Barrett, L., Dunbar, R. and Lycett, J. (2002). *Human Evolutionary Psychology*. Hampshire: Palgrave.

Bateman, A. J. (1948). Intra-sexual selection in Drosophila. *Heredity, 2*, 349-368.

Bateson, P. (1983). *Mate Choice*. Cambridge: Cambridge University Press.

Baulu, J. (1976). Seasonal sex skin colouration and hormonal fluctuations in free-ranging and captive monkeys. *Horm. Behav., 7*, 481-494.

Bercovitch, F. B. (1991). Mate selection, consortship maintenance, and male mating success in savanna baboons. *Primates, 32*, 437-452.

Bercovitch, F. B. (1997). Reproductive strategies of rhesus macaques. *Primates, 38*, 247-263.

Birkhead, T. R. (2000). *Promiscuity. An Evolutionary History of Sperm Competition*. Cambridge, MA: Harvard University Press.

Birkhead, T. R. and Møller, A. P. (1998). *Sperm Competition and Sexual Selection*. London: Academic Press.

Boake, C. R. B. (1989). Repeatability: its role in evolutionary studies of mating behaviour. *Evolutionary Ecology, 3*, 173-182.

Boinski, S. (1987). Mating patterns in squirrel monkeys (*Saimiri oerstedi*). Implications for seasonal sexual dimorphism. *Behav. Ecol. Sociobiol., 21*, 13-21.

Bradbury, J.W. and Andersson, M. B. (eds.) (1987). *Sexual Selection: Testing the Alternatives*. New York: John Wiley and Sons.

Brown, L. (1981). Patterns of female choice in mottled sculpins (*Cottidae, Teleostei*). *Anim. Behav., 29*, 375-382.

Bulmer, M. (1989). Structural Instability of Models of Sexual Selection. *Theoretical Population Biology, 35*, 195-206.

Burk, T. (1982). Evolutionary significance of predation on sexually signalling males. *Florida Entomol., 65*, 90-104.

Butlin, R. K., Woodhatch, C. W. and Hewitt, G. M. (1987). Male spermatophore investment increases female fecundity in a grasshopper. *Evolution, 41*, 221-225.

Carlson, A. (1989). Courtship feeding and clutch size in red-backed shrikes (*Lanius collurio*) *Am. Nat. 133*, 454-457.

Charalambous, M., Butlin, R. K. and Hewitt, G. M. (1994). Genetic variation in song and female song preference in the grasshopper *Chorthippus brunneus* (Orthoptera: Acrididae). *Animal Behaviour, 47*, 399-411.

Clayton, D. H. (1990). Mate Choice in Experimentally Parasitized Rock Doves: Lousy Males Lose. *Amer. Zool, 30 (2)*, 251-262.

Clutton-Brock, T. H. (2007). Sexual Selection in Males and Females. *Science, 318*, 1882-1885.

Clutton-Brock, T. H. and Parker, G. A. (1992). Potential Reproductive Rates and the Operation of Sexual Selection. *The Quarterly Review of Biology, 67 (4)*, 437-456.

Collias, N. E., and Collias, E. C. (1984). *Nest building and bird behaviour.* Cambridge, Mass.: Harvard University Press.

Cords, M. (1987). Forest guenons and patas monkeys: Male-male competition in one-male groups. In Smuts, B.B., Cheney, D. L., Seyfarth, R. M., Wrangham, R. W., and Struhsaker, T. T. (eds.), *Primate Societies*, Chicago. University of Chicago Press, pp.98-111.

Darwin, C. (1871). *The descent of man and selection in relation to sex.* London: Murray.

De Winter, A. J. (1992). The genetic basis and evolution of acoustic mate recognition signals in a *Ribautodelphax* planthopper (Homoptera, Delphacidae). I. The female call. *Journal of Evolutionary Biology, 5*, 249-265.

Dixson, A. F. (1998). *Primate Sexuality: Comparative Studies of the Prosimians, Monkeys, Apes and Human Beings.* Oxford: Oxford University Press.

Dunbar, R. I. M. (1984). *Reproductive decisions: an economic analysis of gelada baboon social strategies.* Princeton University Press.

Eaton, G. G. (1973). Social and endocrine determinants of sexual behaviour in simian and prosiminan females. In, *Symp. IVth Int. Congr. Priamtol., Vol. 2.* (eds. C. H. Phoenix), pp.20-35. Karger, Basel.

Eberhard, W. G. (1996). *Female Control: Sexual Selection by Cryptic Female Choice.* Princeton, NJ: Princeton University Press.

Emlen, S. T. and Oring, L. W. (1977). Ecology, Sexual Selection, and the Evolution of Mating Systems. *Science, 197*, 215-223.

Enomoto, T. (1974). The sexual behaviour of Japanese macaques. *J. Hum. Evol., 3*, 351-372.

Enquist, M., and Arak, A. (1994). Symmetry, beauty, and evolution. *Nature, 372*, 169–172.

Enquist, M., and Johnstone, R. A. (1997). Generalization and the evolution of symmetry preferences. *Proc. R. Soc. Lond. B Biol. Sci. 264*, 1345–1348.

Fisher, R. A. (1930). *The Genetical Theory of Natural Selection.* Oxford: Clarendon Press.

Folstad, I. and Karter, A. J. (1992). Parasites, bright males and the immunocompetence handicap. *Am. Nat., 139*, 603–622.

Gartlan, J. S. and Brain, C. K. (1968). Ecology and social variability in *Cercopithecus aethiops* and *C. mitis.* In *Primates, studies in adaptation and variability* (ed. P. C. Jay), pp. 253-292. New York: Holt, Rinehart and Winston.

Geary, D. C., Vigil, J. and Byrd-Craven, J. (2004). Evolution of Human Mate Choice. *The Journal of Sex Research, 41 (1)*, 27-42.

Goodall, J. (1986). *The chimpanzees of Gombe: patterns of behaviour.* Harvard: Belknap Press.

Grafen, A. (1990a). Sexual Selection Unhandicapped by the Fisher Process. *J. Theor. Biol., 144*, 473-516.

Grafen, A. (1990b). Biological Signals as Handicaps. *J. Theor. Biol., 144*, 517-546.

Gross, M. R. (1994). The evolution of behavioural ecology. *Trends Ecol. Evol., 9*, 358-360.

Gwynne, D. T. (1981). Sexual difference theory: Mormon crickets show role reversal in mate choice. *Science, 213*, 779-780.

Halliday, T. R. (1976). The libidinous newt. An analysis of variations in the sexual behaviour of the male smooth newt, *Triturus vulgaris. Anim. Behav. 24*, 398-414.

Hamilton, W. D. and Zuk, M. (1982). Heritable true fitness and bright birds: a role for parasites? *Science 218*, 384-387.

Harvey, P. H., and Bradbury, J. W. (1991). Sexual Selection. In J. R. Krebs and N. B. Davies (eds.) *Behavioural ecology* (3rd ed), pp. 203-223. Oxford: Blackwell Scientific.

Hillgarth, N. (1990). Parasites and Female Choice in the Ring-necked Pheasant. *Amer. Zool., 30*, 227-233

Hoekstra, R. F. (1987). The evolution of sexes. In S. C. Stearns, (ed.) *The Evolution of Sex and Its Consequences*, pp. 59-92. Birkhauser: Basel.

Holm, C. H. (1973). Breeding sex ratios, territoriality and reproductive success in the red-winged blackbird (*Agelaius phoeniceus*). *Ecology, 54*, 356-365.

Houde, A. E. (1994). Effect of artificial selection on male colour patterns on mating preference of female guppies. *Proceedings of the Royal Society of London, Series B, 256*, 125-130.

Houde, A. E. and Torrio, A. J. (1992). Effect of parasitic infection on male color pattern and female choice in guppies. *Behav. Ecol., 3*, 346-351.

Iwasa, Y. and Pomiankowski, A. (1994). The Evolution of Mate Preferences for Multiple Sexual Ornaments. *Evolution, 48 (3)*, 853-867.

Iwasa, Y., Pomiankowski, A. and Nee, S. (1991). The evolution of costly mate preferences. II. The handicap principle. *Evolution, 45*, 1431-1442.

Janson, C. (1984). Female choice and the mating system of the brown capuchin monkey, Cebus paella (Primates: Cebidae). *Z. Tierpsychol., 65*, 177-200.

Jennions, M. D. and Petrie (1997). Variation in Mate Choice and mating Preferences: A Review of Causes and Consequences. *Biol. Rev., 72*, 283-327.

Johnson, K., Thornhill, R., Ligon, J. D. and Zuk, M. (1993). The direction of mothers' and daughters' preferences and the heritability of male ornaments in red jungle fowl (*Gallus gallus*). *Behavioural Ecology, 4*, 254-259.

Johnstone, R. A. (1995). Sexual selection, honest advertising and the handicap principle: reviewing the evidence. *Biological Reviews of the Cambridge Philosophical Society, 70*, 1-65.

Jones, G. P. (1981). Spawning-site choice by female *Pseudolabrus celidotus* (Pisces: Labridae) and its influence on the mating system. *Behav. Ecol. Sociobiol., 8*, 129-142.

Kaneshiro, K. Y. (1989). The dynamics of sexual selection and founder effects in species formation. In *Genetics, Speciation and the Founder Principle* (ed. L. V. Giddings, K. Y. Kaneshiro and W. W. Anderson), pp. 279-296. New York. Oxford University Press.

Kappeler, P. and van Schaik, C. (2004). *Sexual selection in Primates: New and comparative perspectives*. Cambridge: Cambridge University Press.

Kearns, P. W. E., Tomlinson, I. P. M., Veltman, C. J. and O'Donald, P. (1992). Non-random mating in *Adalia bipunctata* (the two-spot ladybird). II. Further tests for female mating preferences. *Heredity, 68*, 385-389.

Keddy, A. C. (1986). Female choice in vervet monkeys (*Cercopithecus aethiops*). *Am. J. Priamtol., 10*, 125-143.

Keddy-Hector, A. C. (1992). Mate Choice in Non-Human Primates. *Amer. Zool., 32*, 67-70.

Keddy-Hector, A. C., Seyfarth, R. M. and Raleigh, M. J. (1989). Male parental care, female choice and the effect of an audience in vervet monkeys. *Anim. Behav., 38*, 262-271.

Kennedy, C. E. J., Endler, J. A., Poyton, S. L. and McMinn, H. (1987). Parasite load predicts mate choice in guppies. *Behav. Ecol. Sociobiol., 21*, 291-295.

Kirkpatrick, M. (1985). Evolution of Female Choice and Male Parental Investment in Polygnous Species: The Demise of the 'Sexy Son'. *Am. Nat., 125 (6)*, 788-810.

Kirkpatrick, M. (1996). Good Genes and Direct Selection in the Evolution of Mating Preferences. *Evolution, 50 (6)*, 2125-2140.

Kirkpatrick, M. and Ryan, M. J. (1991). The evolution of mating preferences and the paradox of the lek. *Nature, 350*, 33-38.

Kitchen, D. W. (1974). Social behaviour and ecology of the pronghorn. *Wildl. Monogr., 38*, 1-96.

Koepfer, H.R. (1987). Selection for sexual isolation between geographic forms of *Drosophila mojavensis*. II. Effects of selection on mating preferences and propensity. *Evolution, 41*, 1409-1413.

Kokko, H., Brooks, R., Jennions, M. D. and Morley, J. (2003). The evolution of mate choice and mating biases. *Proc. R. Soc. Lond. B., 270*, 653-664.

Krebs, J. R. and Davies, N. B. (1997). *Behavioural Ecology: An Evolutionary Approach.* (4th Ed). Oxford: Blackwell Publishing.

Kuester, J., Paul, A., and Arnemann, J. (1994). Kinship, familiarity and mating avoidance in barbary macaques, *Macaca sylvanus*. *Anim. Behav, 48*, 1183-1194.

Lande, R. (1981). Models of speciation by sexual selection of polygenic traits. *Proceedings of the National Academy of Sciences, U.S.A. 78*, 3721-3725.

Lande, R. (1982). Rapid Origin of Sexual Isolation and Character Divergence in a Cline. *Evolution, 36 (2)*, 213-223.

Little, A. C., Burt, D. M., Penton-Voak, I. S., and Perrett, D. I. (2001). Self-perceived attractiveness influences human female preferences for sexual dimorphism and symmetry in male faces. *Proc. R. Soc. Lond. B Biol. Sci. 268*, 39–44.

Loehle, C. (1997). The pathogen transmission avoidance theory of sexual selection. *Ecol. Model. 103*, 231–250.

Madsen, J. D., Shine, R., Loman, J. and Hakansson, T. (1992). Why do female adders copulate so frequently? *Nature, 355*, 440-441.

Manning, J. T. (1995). Fluctuating asymmetry and body weight in men and women: implications for sexual selection. *Ethology and Sociobiology, 16*, 143-153.

Manson, J. H. (1992). Measuring female mate choice in Cayo Santiago rhesus macaques. *Anim. Behav., 61*, 53-64.

Manson, J. H. (1994a). Male aggression: a cost of female mate choice in Cayo Santiago rhesus macaques. *Anim. Behav. 48*, 473–475.

Manson, J. H. (1994b). Mating patterns, mate choice, and birth season heterosexual relationships in free-ranging rhesus macaques. *Primates 35*, 417–433.

Manson, J. H. (1995). Do female rhesus monkeys choose novel males? *Am. J. Primatol., 37*, 285-296.

Markow, T. A. (1988). *Drosophila* males provide a material contribution to offspring sired by other males. *Func. Ecol., 2*, 77-79.

Maynard Smith, J. (1956). Fertility, mating behaviour and sexual selection in *Drosophila subobscura*. *J. Genet., 54*, 261-279.

Maynard Smith, J. (1991). Models of sexual selection. *Trends in Ecology and Evolution, 6*, 146-148.

Majerus, M. E. N., O'Donald, P. and Weir, J. (1982). Female mating preference is genetic. *Nature, London, 300*, 521-523.

McMinn, H. (1990). Effects of the Nematode Parasite (*Camallanus cotti*) on Sexual and Non-Sexual Behaviors in the Guppy (Poecilia reticulata). *Amer. Zool, 30 (2)*, 245-249.

Mealey, L. M., Bridgstock, R., and Townsend, G. C. (1999). Symmetry and perceived facial attractiveness: a monozygotic co-twin comparison. *J. Pers. Soc. Psychol., 76*, 151–158.

Missakian, E. (1973). Genealogical mating activity in free-ranging groups of rhesus monkeys (*Macaca mulatta*) on Cayo Santiago. *Behaviour, 45*, 224-240.

Modahl, K. B. and Eaton, G. G. (1977). Display behaviour in a confined troop of Japanese macaques. *Anim. Behav., 25*, 525–535.

Møller, A. P. (1988). Female choice selects for male sexual tail ornaments in the monogamous swallow. *Nature Lond., 332*, 640-642.

Møller, A. P. (1992). Parasites differentially increase the degree of fluctuating asymmetry in secondary sexual characters. *J. Evol. Biol, 5*, 691-699.

Møller, A. P. (1990). Effects of a haematophagous mite on the barn swallow (*Hirundo rustica*): a test of the Hamilton-Zuk hypothesis. *Evolution, 44*, 771-784.

Møller, A. P. (1994a). *Sexual Selection and the Barn Swallow*. Oxford. Oxford University Press.

Møller, A. P. (1994b). Symmetrical male sexual ornaments, paternal care, and offspring quality. *Behavioural Ecology, 5*, 188-194.

Møller, A. P. (1994c). Male ornament size as a reliable cue to enhanced offspring viability in the barn swallow. *Proc. Nat. Acad. Sci. USA, 91*, 6926-6932.

Møller, A. P. and Höglund, J. (1991). Patterns of fluctuating asymmetry in avian feather ornaments: implications for models of sexual selection. *Proc. Roy. Soc. Lond. B, 245*, 1-5.

Møller, A. P. (2000). *The Mating mind. How sexual Choice Shaped the Evolution of Human Nature*. New York: Doubleday.

Moore, A. J. (1989). Sexual selection in *Nauphoeta cinerea*: inherited mating preferences? *Behavioural Genetics, 19*, 717-724.

Moore, A. J. (1990). The Evolution of Sexual Dimorphism by Sexual Selection: The Separate Effects of Intrasexual Selection and Intersexual Selection. *Evolution, 44 (2)*, 31-35.

Muldal, A. M., Moffatt, J. D. and Robertson, R. J. (1986). Parental care of nestlings by male red-winged blackbirds. *Behav. Ecol. Sociobiol., 19*, 105-114.

Nakatusuru, K., and Kramer, D. L. (1982). Is sperm cheap? Limited male fertility and female choice in the lemon tetra (Pisces, Characidae). *Science, 216*, 753-755.

Nicoletto, P. F. (1995). Offspring quality and female choice in the guppy. *Poecillia reticulata. Animal Behaviour, 49*, 377-387.

Norris, K. (1993). Heritable variation in a plumage indicator of viability in male great tits *Parus major. Nature, 362*, 537-539.

Nunn, C. L., Gittleman, J. L. and Antonovics, J. (2000). Promiscuity and the primate immune system. *Science, 290*, 1168–1170.

Olsson, M. (1992). *Sexual Selection and Reproductive Strategies in the Sand Lizard (Lacerta agilis)*. Ph.D. diss., University of Göteborg, Sweden.

Packer, C. (1979). Inter-group transfer and inbreeding avoidance in *Papio Anubis. Anim. Behav., 27*, 37-45.

Palombit, R. A., Seyfarth, R. M., and Cheney, D. L. (1997). The adaptive value of "friendships" to female baboons: Experimental and observational evidence. *Anim. Behav., 54*, 599-614.

Parker, G. A. (1970). Sperm competition and its evolutionary consequences in the insects. *Biol. Rev., 45*, 525-567.

Parker, G. A. (1983). Mate quality and mating decisions. In P. Bateson (ed.) *Mate Choice*, pp. 141-166. Cambridge: Cambridge University Press.

Paul, A. (2001). Sexual Selection and Mate Choice. *International Journal of Primatology, 23 (4)*, 877-904.

Perloe, S. I. (1992). Male mating competition, female choice and dominance in a free-ranging group of Japanese macaques. *Primates., 33*, 289-304.

Petrie, M. (1983). Female moorhens compete for small fat males. *Science 220*, 413-415.

Petrie, M. (1994). Improved growth and survival of offspring of peacocks with more elaborate trains. *Nature, 371*, 598-599.

Pomiankowski, A. (1987). The Costs of Choice in Sexual Selection. *J. Theor. Biol, 128*, 195-218.

Pomiankowski, A. Iwasa, Y. and Nee, S. (1991). The Evolution of Costly Mate Preferences I. Fisher and Biased Mutation. *Evolution, 45 (6)*, 1422-1430.

Pomiankowski, A. and Sheridan, L. (1994). Female choice and genetic correlations. *Trends in Ecology and Evolution, 9*, 343.

Raleigh, M. J., D. Pollack, M. T. McGuire, and A. C. Keddy-Hector. (1992). Females prefer males who care for the infants of other females: Implications for choice in vervet monkeys. Cited in Keddy-Hector, A. C. (1992). Mate Choice in Non-Human Primates. *Amer. Zool., 32*, 67-70.

Ralls, K. and Ballou, J. (1982). Effects of inbreeding on infant mortality in captive primates. *Int. J. Primatol., 3*, 491-505.

Ritchie, M.G. (1992). Setbacks in the search for mate-preference genes. *Trends in Ecology and Evolution, 7*, 328-329.

Robertson, J. G. M. (1990). Female choice increases fertilization success in the Australian frog *uperoleia laevigata. Anim. Behav. 39*, 639-645.

Roeder, K. D. (1935). An Experimental Analysis of the Sexual Behavior of the Praying Mantis (*Mantis religiosa L.*). *Biol. Bull., 69*, 203-220.

Roelofs, W. L., Du, J. W., Linn, C., Glover, T. J. & Bjostad, L. B. (1986). The potential for genetic manipulation of the redbanded leafroller moth sex pheromone blend. In *Evolutionary Genetics of Invertebrate Behaviour: Progress and Prospects* (ed. M. D. Huttell), pp. 263- 272. New York: Plenum Press.

Ross, C. (1998). Primate life histories. *Evolutionary Anthropology, 6*, 54-63.

Rutowski, R. L. (1979). Courtship behaviour of the checkered white, *Pieris protodice* (Pieridae). *J. Lepid. Soc. 33*, 42-49

Ryan, M. J. and Keddy-Hector, A. (1992). Directional patterns of female mate choice and the role of sensory biases. *American Naturalist, 139*, 4-35.

Sanderson, N. (1989). Can Gene Flow Prevent Reinforcement? *Evolution, 43 (6)*, 1223-1235.

Severinghaus, L., Kurtak, B. H., and Eickwort, G. C. (1981). The reproductive behaviour of *Anthidium manicatum* (Hymenoptera: Megachilidae) and the significance of size for territorial males. *Behav. Ecol. Sociobiol., 9*, 51-58.

Simmons, L. W. (1990). Post-copulatory guarding, female choice and the levels of gregarine infections in the field cricket, *Gryllus bimaculatus*. *Behav. Ecol. Sociobiol. 26*, 403-409.

Small, M. F. (1989). Female choice in nonhuman primates. *Yearbook of Physical Anthropology, 32*, 103-127.

Smith, D. G. (1994). Male dominance and reproductive success in a captive group of rhesus macaques (*Macaca mulatta*). *Behaviour, 129*, 225-242.

Smuts, B. B. (1985). *Sex and friendship in Baboons*. New York: Aldine.

Soltis, J., Mitsunaga, F., Shimizu, K., Yanagihara, Y. and Nozaki, M. (1999). Female Mating Stategy in an Enclosed Group of Japanese Macaques. *American Journal of Priamtology, 47,* 263-278.

Soltis, J., Mitsunaga, F., Shimzu, K., Yanagihara, Y., and Nozaki, M. (1997). Sexual selection in Japanese macaques I: Female mate choice or male sexual coercion? *Anim. Behav. 54*, 725-736.

Soulé, M. (1982). Allometric variation. I. The theory and some consequences. *Am. Nat., 129*, 751-764.

Tasker, C. R., and Mills, J. A. (1981). A functional analysis of courtship feeding in the red-billed gull, *Larus novaehollindiae scopulinus*. *Behaviour, 77*, 222-241.

Thornhill, R. (1983). Cryptic female choice and its implications in the scorpionfly *Harpobittacus nigriceps*. *Am. Nat., 122*, 765-788.

Thornhill, R. (1992). Female Preference for the pheromone of males with low fluctuating asymmetry in the Japanese scorpion fly (*Panorpa japonica*: Mecoptera). *Behav. Ecol., 3*, 277-283.

Thornhill, R., and Alcock, J. (1983). *The Evolution of Insect Mating Systems*. Cambridge, Mass: Harvard University Press.

Trivers, R. L. (1972). Parental investment and sexual selection. In B. Campbell (ed.) *Sexual Selection and the Descent of Man*, 139-179. Chicago: Aldine.

Van Schaik, C. P., and van Hooff, J. A. R. A. M. (1996). Toward an understanding of the orangutan's social system. In, McGrew, W. C., Marchant, L. F., and Nishida, T. (eds.), *Great Ape Societies*, pp. 3–15. Cambridge: Cambridge University Press.

Van Valen, L. A. (1962). A study of fluctuating asymmetry. *Evolution, 16*, 125–142.

Waitt, C. and Little, A. C. (2006). Preferences for Symmetry in Conspecific Facial Shape Among *Macaca mulatta*. *International Journal of Primatology, 27 (1)*, 133-145.

Waitt, C., Little, A. C., Wolfensohn, S., Honess, P., Brown, A. P., Buchanan-Smith, H. M., and Perrett, D. I. (2003). Evidence from rhesus macaques suggests that male coloration plays a role in female primate mate choice. *Proc. R. Soc. Lond. B., 270*, S144-S146.

Watson, P. J. and Thornhill, R. (1994). Fluctuating asymmetry and sexual selection. *Trends in Ecology and Evolution, 9*, 21-25.

Weatherhead, P. J. and Robertson, R. J. (1979). Offspring quality and the polygyny threshold: "the sexy son hypothesis". *Am. Nat., 113*, 201-208.

Welker, C. Hohmann, H., and Shafer-Witt, C. (1990). Significance of kin relations and individual preferences in the social behaviour of *Cebus paella*. *Folia Primatol., 54*, 166-170.

Wiggins, D. A., and Morris, R. D. (1986). Criteria for female choice of mates: Courtship feeding and parental care in the common tern. *Am. Nat. 128*, 126-129.

Williams, G. C. (1992). *Natural Selection: Domains, Levels, and Challenges*. Oxford: Oxford University Press.

Yerkes, R. M. (1939). Sexual behavior in the chimpanzee. *Hum. Biol., 11*, 78-110.

Zahavi, A. (1975). Mate selection: a selection for a handicap. *Journal of Theoretical Biology, 53*, 205-214.

Zahavi, A. (1977). The cost of honesty (further remarks on the handicap principle). *Journal of Theoretical Biology, 67*, 603-605

Zahavi, A. and Zahavi A. (1997). *The Handicap Principle: a Missing Part of Darwin's Puzzle*. Oxford: Oxford University Press.

Zuk, M., Thornhill, R., Ligon, J. D. Johnson, K. (1990). Parasites and mate choice in red jungle fowl. *Amer. Zool., 30*, 235-244.

In: Animal Behavior: New Research
Editors: E. A. Weber, L. H. Krause

ISBN: 978-1-60456-782-3
© 2008 Nova Science Publishers, Inc.

Chapter 2

FEMALE MATE CHOICE IN NON-HUMAN MAMMALS

Benjamin D. Charlton

Department of Psychology, School of Life Sciences,
University of Sussex, Sussex, England BN1 9QH

ABSTRACT

Until now definite studies of mate choice have typically focused on non-mammal animal species where short life spans and gestation periods make reproductive success quicker to determine, and in which the females show preferences for clearly defined morphological or behavioural male traits. Mammals, which are invariably larger, have longer life spans and inter-birth intervals and are behaviourally more complex, are less suited to this type of experimentation. However, notwithstanding these difficulties, many studies of mate choice have been conducted on mammals and the current literature reveals that the females of several mammal species do appear to choose their mates, and through this potentially gain important direct and indirect fitness benefits. Here I review this body of work to reveal that the majority of female mammal mate choice studies conducted so far, particularly on large mammals, fail to: 1) determine the actual male phenotypic trait (s) of female preference (which is crucial to identifying male characteristics under sexual selection); 2) Take into account other environmental and social factors that may affect female response to male roars, as well as intrinsic factors such as female hormonal state and breeding status; and 3) quantify the fitness benefits to discriminating females. I go on to give suggestions for future research and emphasise the need for a combination of carefully designed experimental and field studies. Experimental setups allow us isolate specific male traits from other aspects of the male phenotype, determine female hormonal state and control for other competing male and female mating strategies that may affect female behaviour. Field observations of female mating behaviour can then be conducted to determine whether behavioural responses reported using experimental setups translate to actual copulations in natural conditions, and hence affect the reproductive success of individuals. Only this integrative approach will allow us to gain an appreciation of how inter-sexual selection is generated on specific aspects of the male phenotype, and ultimately enable us to understand the link between mating preferences, female mate choice and reproductive success in mammals.

INTRODUCTION

Sexual selection theory predicts that female mammals should preferentially mate with high quality males, choosing the sire of their offspring for direct benefits, that increase female survival or fecundity, or indirect (genetic) benefits that increase offspring viability and/or attractiveness (Andersson 1994). Traditionally, the role of female mate choice, especially in polygynous mammals, has been considered negligible compared with male-male competition. However, on the basis of the relatively higher level of maternal to paternal investment and generally lower potential rates of reproduction the females of most mammal species could be expected to select their mating partners (Trivers 1972; Clutton-Brock and Vincent 1991). Although the fitness payoffs for discriminating females (and hence the evolutionary trajectories through which female preferences arise) are still much debated (Kirkpatrick and Ryan 1991; Jennions and Petrie 1997; reviewed by Cordero and Eberhard 2003; Kokko et al. 2003), the adaptive value of female mate choice in mammal species, where males differ in their reproductive value to females, is now generally accepted (Andersson 1994; Paul 2002).

Studies of female choice in mammals are mostly hindered by the confounding influence of other male and female mating strategies. Female mating strategies may fulfill non-procreative roles, for example, multi male mating may reflect the need to confuse paternity in mammals where infanticide is common (Wolff and Macdonald 2004) or to maintain the resources and support of male social group members (Gagneux et al. 1999). In addition, male strategies may limit a female to mating with only one male whatever her preference criteria (Soltis 1999), making it imperative to make the distinction between female mate choice and female mating preferences (Halliday 1983). Interpreting patterns of mate choice in terms of variation in mating preferences is problematic as female mating preferences, referring only to internal motivation, are just one component of female mate choice. Moreover, variation in female mate choice can be due to variability in sampling tactics, which are in turn affected by external factors (risk of predation, male harassment), as well as preference functions (Jennions and Petrie 1997). Hence, in natural conditions mating preferences are expressed in a field of environmental and social constraints imposed on mate searching females. For these reasons, female mate choice is best described "as the differential mating by females as a result of the interaction of environmental conditions, mating preferences and sampling strategies" (Wagner 1998). However, it must also be noted that females may even be passively attracted to a more conspicuous signal and not, after sampling several potential mates, actively choosing an individual at all (Forrest and Raspet 1994). Finally, mate choice is not just restricted to active discrimination between potential mating partners but can also be indirect via any other behaviour that increases or decreases the chances of mating with a specific subset of males (Wiley and Poston 1996). Any female mating strategies that incite male-male competition, e.g. reproductive synchrony (Chism and Rogers 1997) or female promiscuity (Preston et al. 2003), so that only the highest quality males can gain access to females, may represent examples of indirect mate choice.

Female mate choice, whether direct or indirect, is constrained by the environment and can only be realistically observed under natural conditions. However, in natural conditions male-male competition can generate the same assortative mating patterns as would female choice and it becomes hard to separate the two processes of selection (Andersson 1994). To determine female mating preferences appropriate experimental setups are required that control for male-male competition and isolate the male trait of preference from other aspects of the male phenotype. A further complication of mate choice studies lies in showing that male traits of female preference are positively correlated to male reproductive success in natural conditions i.e. that they influence reproductive outcomes, such as the viability and/or attractiveness of the offspring (Searcy and Andersson 1986). To do this requires genetic paternity analysis studies that determine the reproductive success of different male phenotypes and their subsequent offspring.

Hence, to gain a complete understanding of female mate choice in any given species requires a combination of field and experimental studies. Notwithstanding these problems, many studies of mate choice have been conducted on mammals and important insights gained. The current literature reveals that the females of several mammal species do choose their mating partners and through this potentially gain important direct and indirect fitness benefits.

FEMALE MATE CHOICE STUDIES

Choice for Direct Benefits

Evidence that female mammals choose males for resources that directly increase their fecundity or reduce reproductive costs is less abundant than it is in birds (Lampe and Espmark 2003) and insects (Thornhill 1976) but does exist. Female mate choice in some highly social mammals for more familiar males may be based on complex delayed preferences for individuals that have proved themselves better able to provision their females over time, as has been suggested in spotted hyenas (*Crocuta crocuta*) (East and Hofer 1991). However, conjecture aside, few studies actually show that female mammals choose mates for direct benefits or indicate what criteria are selected.

There is some evidence to suggest female choice based on territory quality occurs in certain mammal species. A study on a facultatively monogamous population of Pikas (*Ochotona princeps*) reports female choice of mate based primarily on nest quality and location (Brandt 1989). Females preferred single entrance dens located close to food resources for their nests, factors also positively correlated to reproductive success. In this case, males did not actively interfere with the mating attempts of other individuals and hence females may have been able to exercise relatively free choice. However, the importance of other male phenotypic characteristics not considered in this study, such as vocal displays (that could be used by females to discriminate between males) is unknown. It has also been suggested that the females of some monogamous primates may prefer males that are better able to defend a territory (Smuts 1987). However, it can be difficult in any territorial species to determine whether females are attracted to individual males *per se* or the territory they currently occupy. As only the highest quality males are able to acquire and hold on to the 'preferred' territories, any findings are probably confounded by choice for male phenotype. A further complication in quantifying female choice for territory is that it may simply be the result of females copying the preferences of other individuals (Brooks 1998) or appearing to due to the incidental consequence of some other adaptive behaviour e.g. females congregating together to reduce predation risk or harassment (Clutton-Brock and McComb 1993).

Iberian red deer males (*Cervus elaphus hispanicus*) compete for and defend territories that females are subsequently attracted too (Carranza 1995). Although male phenotype would still appear to be important for attracting females, as some males maintained higher attraction rates than others on the same territories, male reproductive success in this species seems to be largely dependant on the quality and/or position of territory held. Some other studies on polygynous ungulates have suggested female mate choice based at least partly on territory at leks (Apollonio et al. 1990; Balmford et al. 1992a). These mating systems are traditionally thought to provide only indirect benefits to females, however, female choice driven at least in part for direct benefits would go some way to solving what has become known as the 'lek paradox' (Borgia 1979; Reynolds and Gross 1990).

Other studies have suggested that female mammals obtain direct benefits in the form of reduced harassment when selecting to mate with males on leks (Balmford 1991; Clutton-Brock et al. 1992; Nefdt and Thirgood 1997). However, this is also disputed by work reporting higher rates of harassment for females on leks (Bro-Jorgensen 2002). In any case, harassment by other courting males can impose serious fitness costs which mate searching females would be expected to avoid (Reale et al. 1996). It would appear that certain female mating tactics have evolved to reduce male harassment costs (Galimberti et al. 2000) and females may also be expected to have evolved preferences for males with certain characteristics that can reduce disturbances from other males. This appears to be the case in South American sea lions (*Otaria flavescens*: Cassini and Fernandez-Juricic 2003) and possibly Sumatran orangutans (*Pongo pygmaeus abelii*: Fox 2002) where female association with adult males, to reduce harassment costs, may be a factor influencing mate choice. Female attraction to males with higher roaring rates in red deer (*Cervus elaphus*) may also reflect a choice for high quality males that are better able to protect females from harassment and hence a choice based partly on direct benefits to the female (McComb 1991).

In mammals where parental care limits female reproductive success females could be expected to show a choice for males that demonstrate infant care abilities. Some male primates occasionally carry infants and, as carrying may reduce the risks of infanticide and hence be of direct benefit to the female, this could be seen as a form of parental care (Palombit et al. 1997). However, this care may simply be a male mating strategy if females mate preferentially with males that have previously demonstrated infant care (Keddy-Hector 1992). In male lions (*Panthera leo*) mane darkness is correlated to serum testosterone and nutrition, reflecting good fighting ability and short-term health of the bearer (West and Packer 2002). Thus, females could gain direct benefits from preferring males with darker manes that would be more adept at protecting their cubs from infanticide and providing food for the pride. These predictions remain untested and currently there appears to be no documented cases of female choice based on parental care abilities in mammals. However, some monogamous mammals could be expected to exhibit bi-parental care, as certainty of paternity for males would be higher than it is in polygynous mammals. More research is required to determine whether or not this is the case. Lastly, there may also be direct incentives for females to avoid parasitized males in mating systems where the female appears to gain nothing more from than a male's sperm from copulation (Kirkpatrick and Ryan 1991). Effects of parasite infection on mate choice may therefore reflect the avoidance of direct costs, if fecundity is affected by parasite infection, rather than Hamilton-Zuk style indicator systems.

Choice for Indirect (Genetic) Benefits

In mating systems where males contribute little but their genes to their offspring it may be that females are choosing males for the indirect benefits of 'good genes' (Darwin 1871; Maynard Smith 1991). The expression of a specific morphological trait of preference in the offspring whether it advertises 'genetic quality' or is arbitrarily 'attractive' to females will yield higher lifetime reproductive success via increased longevity or mating opportunities for the bearer (Kokko et al. 2002). Female mate choice for male phenotype is undoubtedly important in lekking mammals (Clutton-Brock et al. 1989; Balmford 1991; Bro-Jorgensen 2002) but the exact criteria on which female preferences are based in these mating systems is still fairly unclear. The majority of mate choice studies conducted so far on mammals claim to demonstrate female preferences for dominant individuals. As high status may indicate high genetic quality, this suggests that choice for 'good genes' may be an important selective force in mammals.

A recent study investigating the relationship between male dominance ranks, female mate choice and reproductive success in captive chimpanzees (*Pan troglodytes*) reports female choice for dominant males (Klinkova et al. 2005). However, in this case, the positive correlation between male dominance rank and reproductive success observed would appear to be generated predominantly by rank related differences in female solicitations and competitive ability. Indeed, with a social climate characterized by male coercion, in which females are unlikely to be able to exercise free choice, it may be impossible to determine whether any choice reflects true underlying female preferences. Dominant males will often monopolize access to females during the breeding season or at peak conception time (Soltis 1999; Preston et al. 2003), especially in mating systems where female mating tactics incite male-male competition (Semple 1998; Wolff 1998). Other researchers investigating mammal mate choice in natural conditions, propose that females choose dominant or more vigorous males on the basis that subordinates are rejected: reindeer (*Rangifer tarandus*: Hirotani 1989); bighorn sheep (*Ovis canadensis Canadensis*: Hogg 1987); pronghorn deer (*Antilocapra Americana*: Byers et al. 1994) and bison (*Bison bison*: Wolff 1998) or because female solicitation rates are higher to dominant individuals: chimpanzees (*Pan troglodytes*: Matsumoto-Oda 1999; Vorobieva et al. 2004); Japanese macaques (*Macaca fuscata*: Soltis et al. 1997); ring-tailed mouse lemurs (*Lemur catta*: Parga 2002) and vervet *monkeys (Cercopithecus-Aethiops*: Keddy 1986*)*. However, the findings of the above studies cannot be interpreted as evidence for active female choice of dominant males when in actual fact the female's ability to express preferences is likely to be constrained by the social environment. In addition, avoidance of young or subordinate males by females is common in several mammal species (Cox and Leboeuf 1977; Clutton-Brock et al. 1982; Miura 1984; Hirotani 1989; Komers et al. 1999) and hence, whether females are directly choosing dominant individuals remains speculative.

When other competing males are around there are likely be other reasons for associating with older or more dominant individuals who may be expected to have greater abilities to acquire and defend resources as well as provide increased protection from sexual coercion, harassment or infanticide (Pope 1990; Pereira and Weiss 1991). Indeed, female capuchins cease to prefer the dominant male when he loses his dominance status, suggesting that they are not selecting these individuals for their 'good genes' (Janson 1984). Moreover, dominant male mammals do not always consistently attain high reproductive or even mating success (Takahata et al. 1999). In any case, it is imperative to control for male competition so that female mate choice can be studied free of constraints and mating preferences can be determined.

Experimental Studies of Female Preferences in Mammals

Until now definite experimental studies of mate choice have typically focused on non-mammal animal species e.g. insects (Klappert and Reinhold 2003), frogs (Murphy and Gerhardt 2000) and birds (Mountjoy and Lemon 1996) where short life spans and gestation periods make reproductive success quicker to determine and in which the females show preferences for clearly defined morphological or behavioural male traits. Mammals, which are invariably larger, have longer life spans and inter-birth intervals and are behaviourally more complex, are less suited to this type of experimentation. Notwithstanding these problems, some studies on small mammals reporting direct female mate choice for dominant males have controlled for male competition using laboratory experiments: house mouse (*Mus musculus*: Rolland et al. 2003; Tomihara 2005) and field experiments: golden hamsters (*Mesocricetus auratus*: Lisk et al. 1989); brown lemmings *(Lemmus-Trimucronatus*: Huck and Banks 1982)*; voles (*Microtus-Ochrogaster and Microtus-Montanus*: Shapiro and Dewsbury 1986); bank voles (*Clethrionomys glareolus*: Horne and Ylonen 1996) and brown capuchins (*Cebus-Apella*: Janson 1984; Welker et al. 1990). However in all these cases, the exact male phenotypic trait(s) of female preference are not known.

A recent experiment on captive grey mouse lemurs (*Microcebus murinus*) employed a two-way mate choice design in which females could choose between two separately caged males, hence controlling for male-male competition (Craul et al. 2004). The authors report a consistent spatial and/or behavioural female preference, during oestrus, for the male with the higher trill call activity. However, as the authors report, higher trill rate is also positively correlated to male dominance rank and hence, it is impossible to determine whether females are actually attracted to calls with higher trill rates or if this is simply a by-product of this correlation. In addition, only 4 copulations (out of 12 females) were observed and hence whether these preferences translate to increased mating, or more importantly, reproductive success is unknown i.e. whether mating with males with higher trill call activity actually increases offspring viability. Indeed, an earlier study on the same species of mouse lemur reports that dominant males only sired half the offspring in a captive environment in which they may be expected to be able to monopolise females (Radespiel et al. 2002). Another study on the wild guinea pig (*Galea musteloides*), controlling for male competition, reports female preferences for heavier and more frequently courting males (Hohoff et al. 2003). However, in this case the high degree of promiscuity shown by the females suggests that sperm competition may be of greater importance than pre-copulatory mate choice. In this species, as with several primates, mating with larger more dominant males may fulfil other non-procreative functions and again, evidence that it results in more viable offspring is lacking.

In theory, females may also be expected to mate with older males for genetic benefits because viability selection leads to older males with higher genetic quality than those that are younger (Trivers 1972; Kokko and Lindstrom 1996). This theory is backed up to some extent by studies reporting female preferences for older males in birds and insects (Simmons and Zuk 1992; Grahn and Vonschantz 1994). Although no empirical evidence exists in mammals, older males may also be preferred where age attests to viability in the current environment. In spotted hyenas (*Crocuta crocuta*) it has been suggested that male tenure may serve as an index of intrinsic genetic quality, although male fertility may also decline with extreme old age (East et al. 2003). Indeed, a recent theoretical model by Beck & Powell (2000) suggests that preferences for older males are unlikely to evolve in mating systems in which the male provides only sperm. In any case, empirical evidence for increased offspring viability would be required to determine whether females are actually choosing older males for their 'good genes'. In fact, it appears that the only demonstration of increased progeny viability due to free female mate choice in mammals has been carried out on house mice (*Mus musculus*) (Drickamer et al. 2000). In this study, females that were allowed to choose their mates produced offspring of greater viability than those mated at random. However, although this represents empirical evidence of female mate choice for 'good genes', the actual male phenotypic trait(s) of female preference are again not known.

Indeed, identifying the actual trait of female preference, crucial to identifying male characteristics under sexual selection, can be extremely difficult. In addition, female preferences in many mammals may only emerge at specific times such as oestrus (Poole 1989; Matsumoto-Oda 1999; Pillay 2000) or in particular contexts (Qvarnstrom 2001) and in any case, seem likely to be based on multiple cues to male quality, of different or the same modalities (Candolin 2003). It follows therefore that in order to gain a foothold in this immensely complex subject, we must use appropriate experimental setups that control adequately for all other extrinsic and intrinsic factors that affect female behaviour, and in which specific phenotypic traits are isolated from other aspects of the male phenotype. Female responses to variation in specific signal components can then be assessed and an appreciation of how inter-sexual selection is generated on specific aspects of the male phenotype can be gained.

Female Preferences for Specific Phenotypic Traits Signalling Male Genetic Quality

Preliminary evidence that females of some mammal species may choose males for specific phenotypic traits comes from experimental set ups which are able to isolate specific male traits from other aspects of the male phenotype. However, conclusively demonstrating 'attraction' to an isolated trait variant, and hence inferring that mate choice decisions could be made based on them, is difficult.

Among mammals, primates possess the most brilliant secondary sexual colouration that may serve to advertise the bearer's quality to potential mates. Examples include the brightly coloured faces of male mandrills (*Mandrillus sphinx*) (Setchell and Dixson 2001) and scrotal colour in vervets (*Cercopithecus-Aethiops*) (Gerald 2001). Whether female mandrills or vervet monkeys attend to these colour differences between males, however, is unknown. Rhesus macaque females (*Macaca mulatta*) gaze for longer at computer-manipulated red over pale versions of male faces (Waitt et al. 2003). In this case, male coloration might provide a cue to male quality but to determine whether these visual preferences indicate sexual preferences rather than constituting a more general response requires further research. Female preferences for darker manes have been demonstrated in lions (*Panthera leo*) (West and Packer 2002). Darker manes increase heat load for the male lions and may represent an honest indicator of 'good genes' (handicap theory: Zahavi 1975), as males with darker manes are presumably better able to withstand heat-related costs. Indeed, the study reports that males with the darkest manes had superior survival and competitive abilities resulting in increased offspring fitness, but that mane darkness was not heritable. It appears that male colouration, by indicating genetic quality, may play a part in adaptive female mate choice in mammals. Designing appropriate mate choice experiments to answer this question remains a challenge for the future.

Olfactory cues are relatively simple to isolate and present to females in controlled conditions and some of the best evidence of mate choice for 'good genes' in mammals has been generated in this way. Female mice (*Mus domesticus*) and rats (*Rattus rattus*) appear able to discriminate between parasitized and non-parasitized males based on urine odours (Willis and Poulin 2000; Kavaliers et al. 2003) and display increased stress responses to the urine odours of parasite-infected males (Kavaliers et al. 1998). In addition, female house mice mate preferentially with non-parasitized individuals (Ehman and Scott 2002), possibly securing resistant, and hence, 'good' genes for their offspring (Hamilton and Zuk 1982). Female odour preferences for resident males are also reported in rabbits (*Oryctolagus-Cuniculus*) (Reeceengel 1988) however, if these males actually achieve higher reproductive success in natural conditions remains unknown.

Scent over-marking is a common form of competitive advertisement among many mammal species and some work has suggested that scent frequency and placement may play a role in mate choice, possibly for dominant males that mark over subordinates (Johnston et al. 1997; Johnston and Barot 2004) . The Johnston et al. (1997) study on meadow voles (*Microtus pennsylvanicus*) reports female preferences for familiar over novel males even when their scents have been over marked by novel males; however, the ability of these females to discriminate between either novel or familiar male scent markings placed on top of each other could be disputed. Indeed, studies on prairie voles (*Microtus ochrogaster*) in which mate choice based on these criteria had been hypothesised but subsequently disproved, suggests that scent quality may be more important than scent frequency and placement for mate attraction (Thomas 2002; Mech et al. 2003).

Vocal displays can also be reproduced accurately and isolated from other aspects of the male phenotype using playback experiments. Several previous studies on mammals have suggested that male calls may function directly to advertise male quality and attract females without testing experimentally whether or not this is the case: old world monkeys (Gautier and Gautier 1977); pikas (*Ochotona princeps*) (Conner and Whitworth 1985); hammerhead bats (*Hypsignathus monstrosus*) (Bradbury 1977); and some ungulates (Kiley 1972; McElligott et al. 1999). The structurally complex songs of lone humpback (*Megaptera novaeangliae*) and finback whales (*Balaenoptera physalus*) and the underwater acoustic displays of male harbour seals (*Phoca vitulina*) are also thought to function in broadcasting male quality and attracting females (Tyack 1981; Hanggi and Schusterman 1994; Croll et al. 2002). More detailed work on humpback whales, although confirming that these acoustic displays mediate approach and avoidance response, reports no female attraction to male song (Tyack 1983; Mobley 1988). However, the possibility that the female humpbacks use male vocal displays to assess potential mates, only allowing certain males to approach them, cannot be excluded. Similarly, although the long calls of male orangutans (*Pongo pygmaeus*) were originally thought to function in mate attraction, playback experiments to specifically test this did not yield the predicted results (Mitani 1985). This further highlights how crucial it is to test predictions about mate choice using systematic playback experiments.

Recent work on fur seals (*Arctocephalus spp.*) suggests that the divergence in bark calls (which are only given during male-female interactions) could be due to female choice but not full threat calls directed at other males (Page et al. 2002). Indeed, although female mate choice is very likely to be a significant selection force on calls of many male mammals, currently, the only mate choice studies conducted on any mammal that systematically demonstrate a female preference for a vocal characteristic are in red deer (*Cervus elaphus*) (McComb 1991; Charlton 2006; Charlton et al. 2007b). The McComb (1991) study demonstrated a female preference for male roars delivered at higher rates. As male roaring is energetically costly to produce this may represent an honest signal of genetic quality e.g. a handicap (Zahavi 1975; Grafen 1990) and hence a choice for good genes; however, roaring rate seems more likely to reflect current short-term condition or motivation rather than long-term genetic quality. More recent work using an experimental setup showed that oestrous female red deer preferred male roars in which lower formant frequencies simulated larger caller's (Charlton et al. 2007b). To date, this remains the only female mate choice study in mammals to demonstrate a mating preference for an isolated male phenotypic trait signalling 'good genes' (i.e. larger body size) whilst controlling for female hormonal state and male-male competition.

Choice for Compatible Genes

Another aspect of female choice for male genetic 'quality' concerns genetic compatibility (Trivers 1972; Neff and Pitcher 2005) with an increasing body of empirical studies to suggest that females should benefit, by increasing offspring fitness, through choosing genetically dissimilar males (Mays and Hill 2004). Indeed, choice for unfamiliar or novel partners as suggested in several mammals may function to increase offspring genetic heterozygosity and subsequent fitness (Manson 1995; Amos et al. 2001). However, in these cases females should choose in a relative rather than an absolute sense and prefer males with genotypes that best complement their own rather than for 'good genes' *per se*. This method of mate selection does not lead to genetic fixation of male traits, a problem for good genes and runaway selection models, as females should not be congruent in their choice for particular male phenotype i.e. a male that is best for one female may not be best for another (Brown 1997). Choice for compatible genes may have evolved to avoid the reproductive costs associated with inbreeding and distant outbreeding. Female white-footed mice (*Peromyscus leucopus*) gain maximum reproductive success when mated with males of intermediate genetic similarity (Keane 1990) and may represent an example of this type of mate choice, but how the mating preferences observed in this study are mediated is unclear. Indeed, the major problem facing the theory of female choice based on genetic compatibility is in how females actually recognise genetically incompatible mates.

There is good evidence to suggest that the genes of the major histocompatibility complex (MHC), that play a key role in the immune system, function as cues to mate choice for genetically compatible males (reviewed in Penn and Potts 1999). Generally, the mating preference seems to be MHC-disassortative i.e. males with dissimilar major histocompatibility complex alleles are preferred (Penn 2002). This mating pattern leads to more outbred and heterozygous offspring, which has been suggested to increase offspring fitness (Brown and Eklund 1994; Grob et al. 1998). Indirect evidence for MHC mediated inbreeding avoidance comes from Potts et al. (1991) who, having found significantly fewer MHC-homozygous offspring than expected, suggested that female mice *(Mus domesticus)* sought extra-territorial copulations with MHC dissimilar males. However, the results of this study may be the product of simple non-MHC mediated inbreeding avoidance. Convincing evidence for MHC-disassortative mating preferences in mammals has been generated by work on captive mice *(Mus domesticus)* where discrimination between different MHC genotypes seems to be mediated through specific odour cues (reviewed by:Jordan and Bruford 1998). MHC-disassortative mating preferences may increase the resistance of offspring to infectious diseases (Penn 2002) and/or enable individuals to recognize closely related kin and function to avoid inbreeding (Yamazaki et al. 2000). Since MHC genes influence resistance to many infectious diseases, females who prefer disease resistant males may confer these resistant alleles on their offspring (Hamilton and Zuk 1982), however, it remains to be shown whether males with attractive displays carry resistant MHC alleles. Indeed, it is not clear whether the main function of MHC-disassortative mating preferences is enhanced progeny immunocompetance or inbreeding avoidance. Either way, there are potential indirect fitness benefits available to discriminating females.

MHC-dependant preferences may exist across several mammal species. For example, MHC variation is related to antler development and body mass in white tailed deer *(Odocoileus virginianus)* and hence indicates 'good genes' (Ditchkoff et al. 2001). Interestingly, although MHC variation in soay sheep *(Ovis aries L.)* is associated with resistance to a parasitic nematode (Paterson et al. 1998) no evidence for MHC-disassortative mating preferences exists in this species (Paterson and Pemberton 1997). In this case, any female preference may have been masked by intense male-male competition but nevertheless demonstrates the need to test these predictions. Theoretically, female preferences for common MHC alleles, and hence genetic similarity at these loci, could also arise where an evolutionary 'arms race' type scenario exists between hosts and parasites that have not yet adapted to recently common alleles (Hamilton and Zuk 1982). Female striped mice *(Rhabdomys pumilio)* prefer 'homotype' males from their own population than 'heterotype' males, a preference for genetically similar males possibly mediated by MHC alleles (Pillay 2000). However, inbreeding avoidance would appear to offer the highest potential fitness benefits to females (Penn 2002). A recent study on wild house mice *(Mus domesticus)* reported that inbred males sired only one fifth as many surviving offspring as outbred males (Meagher et al. 2000). Female choice for unfamiliar or novel partners is observed in several mammals (Manson 1995) and may in some cases be mediated by MHC gene complexes in mammal species at risk of inbreeding (Clarke and Faulkes 1999; Valsecchi et al. 2002). While there is much evidence that MHC-dependant mating preferences exist in house mice further research is required to determine their generality to other mammal species.

Post-copulatory Choice

Post-copulatory or 'cryptic' mate choice, in which females selectively abort the sperm or offspring of certain males, can take two forms; directional post-copulatory choice in which females bias sperm use after copulation towards certain male phenotypes and non-directional in which females favour the sperm of males with compatible genotypes regardless of their phenotype.

The copulation calls of some female primates that, by encouraging mate guarding and minimizing sperm competition, increase the chances of fertilization by preferred males may represent an example of directional post-copulatory mate choice in mammals (Maestripieri and Roney 2005). This is a compelling argument and contrary to previous suggestions that these copulation calls function to increase sperm competition (Oconnell and Cowlishaw 1994; Semple 1998). Indeed, an earlier study on barbary macaques (*Macaca sylvanus*) also reported increased mate guarding by consort males in response to copulation calls, which again is consistent with the post copulatory choice hypothesis (Todt et al. 1995).

Non-directional post-copulatory mate choice can take place via mechanisms that have evolved to prevent females from being fertilized by incompatible sperm (Zeh and Zeh 1996; Zeh and Zeh 1997). There is evidence in mice (*Mus domesticus*) that MHC-derived proteins expressed on sperm may influence the likelihood that a given spermatozoan will fertilise the egg, with MHC-similar pregnancies spontaneously aborted (Rulicke et al. 1998). Conversely, female grey mouse lemurs (*Microcebus murinus*) actively delay pregnancies more when housed with unfamiliar males suggesting post-copulatory choice for familiar males (Radespiel and Zimmermann 2003). Whatever the preference, post-copulatory mate choice may represent a powerful selective force, especially in mating systems dominated by male coercion and in which females are unlikely to be able to exercise pre-copulatory mate choice.

CONCLUSION

Female mate choice in mammals represents a challenging area of research with many questions still to be answered. While it appears that female mammals are provided with important direct and indirect benefits through mate choice, the functions and evolutionary consequences of such choice remain unclear. To answer these questions, definite evidence of how mate choice for particular male phenotypes impacts on female fecundity and reproductive success is required. If direct benefits are available, they are likely to have a much greater effect on female fitness than indirect benefits, and selection for males providing such benefits is easy to understand (Price et al. 1993). Surprisingly, although direct benefits may provide a possible explanation for some female mammal mating tactics (Wolff and Macdonald 2004), empirical evidence for mating preferences based on them is distinctly lacking in mammals, although direct fecundity benefits may underlie some mating preferences on leks. This apparent omission may arise because the majority of mate choice studies so far have been conducted on polygynous mammals in which males appear to provide little, if any, direct benefits to females and subsequently where female preferences based on them would be harder to observe. Mate choice for indirect benefits, because it is generally assumed that it will result from a genetic association between male attractiveness and offspring fitness, has always been controversial and considered insufficient to explain the evolution of female mating preferences alone (Kirkpatrick 1996). However, a recent theoretical model by Moller & Jennions (2001) suggests that direct benefits are only slightly more important than indirect ones arising from female choice. In reality, female mate choice in most mammal species will almost certainly be influenced to some extent by both direct and indirect benefits. A study on puku (*Kobus vardoni*) and topi antelopes (*Damaliscus lunatus*) in which male phenotype, predation risk and territory quality were all related to female presence appears to confirm this (Balmford et al. 1992b). However, because putative mate choice cues are often interrelated, their relative importance for mate choice becomes difficult to ascertain.

A major problem with the majority of mammal studies is in how female preferences are quantified. Many experimental studies rely on indirect methods to assay female preferences such as increased attention towards an isolated male trait, which does not necessarily indicate attraction. For example, greater attention by red deer hinds to roars simulating males of sub-adult body size than to those simulating a large adult male, whilst making no sense from a mate choice perspective, is likely to indicate an aversive response to facilitate earlier detection and avoidance of young stags that are known to harass them (Charlton 2006; Charlton et al. In submission). Observational studies often use high levels of interaction with certain individuals, which may be due to social and/or environmental constraints or for non-procreative reasons, to suggest a mating preference. In addition, many of these studies only record mating success, as individuals gaining copulations or proximity to members of the opposite sex, and do not necessarily record the reproductive success of individuals. It is essential to conduct genetic paternity analysis studies, to determine the reproductive success of trait bearing individuals and their offspring in natural environments, and hence determine any reproductive benefits to choice. Moreover, realistic assessments of the costs of female mate choice need to be carried out under approximately natural conditions when food supplies are limited and predators and competitors are present. These limitations on obtaining accurate measures of the costs and benefits of mate choice, as well as interacting and often competing male and female mating strategies, make it difficult to test theoretical predictions on mammals.

Some theoretical models of female mate choice behaviour suggest that females should be flexible in their mate choice decisions (Real 1990; for a review see Jennions and Petrie 1997) and recent empirical data shows plasticity of female preferences in animals (Moore and Moore 2001; Veen et al. 2001; Welch 2003; Chaine and Lyon 2008) even during a single reproductive cycle (Lynch et al. 2005). Studies of female choice in mammals, in which female preferences may only emerge during the period of peak conception, also emphasize the importance of considering the timing of female ovulation (Matsumoto-Oda 1999; Stumpf and Boesch 2005). Recent work on oestrous female red deer demonstrated a mating preference for male roars simulating larger caller's (Charlton et al. 2007b), a trait positively correlated to male reproductive success in this species (Reby and McComb 2003). Importantly, by using an experimental setup, this study not only controlled for female hormonal state but was also able to remove other environmental and social factors that may otherwise obscure female mating preferences. Indeed, studies on other animals without observable female mate choice in natural conditions have illustrated that otherwise hidden female preferences do in fact exist and emerge under appropriate experimental conditions (Gould et al. 1999).

Future work should follow this type of experimental approach and concentrate on demonstrating female preferences for specific male traits using setups that control for the effects of male and female mating strategies and female hormonal state. Experimental investigation of female preferences, because they are highly dependant on female hormonal state, should be conducted during and outside peak conception times. Comparisons of female reproductive behaviour during and outside peak conception times has yielded interesting results in tungara frogs *(Physalaemus pustulosus)* (Lynch et al. 2005) and house mice *(Mus musculus)* (Rolland et al. 2003) and may similarly provide important insights into female mate choice in large mammals. Female preferences are also likely to vary geographically due to evolved genetic/environmental differences (Qvarnstrom 2001). Therefore, interesting comparisons could also be made of the female mating preferences of a given mammal species in different environmental conditions. Moreover, experiments using repeated measures designs would provide more complete information about the form of female preferences in mammals, enabling preference functions (how the strength of a female preference varies according to the expression of the male trait) to be built up for individual females as well as for populations (Wagner 1998). In addition, as mate choice studies in general have paid less attention to variation between individual females than males (Jennions and Petrie 1997) the examination of both within and between-female variation in preferences may prove to be a productive area of research in mammals.

Mate-searching females are also likely to assess multiple cues when making mate choice decisions (Candolin 2003; van Doorn and Weissing 2004), yet no mammal mate choice studies have explicitly examined the relative importance of different male traits to female choice. To do this would require experiments in which different male phenotypic trait values within a stimulus are independently varied and presented to females in order to ascertain their relative importance for mate choice. In particular, this approach would allow us to examine how male traits interact to influence female mating decisions. Such interactions undoubtedly have important implications for sexual display evolution in animals' generally e.g. satin bowerbirds *(Ptilonorhynchus violaceus:* Patricelli et al. 2003) and appear to have important effects on female mate choice behaviour in some anuran *(Hyla gratiosa:* Poole and Murphy 2007) and insect species *(Gryllus campestris:* Scheuber et al. 2004), but as yet represent an unexplored avenue of research in mammals. A recent study on red deer independently manipulated components of the male roar, the fundamental frequency and the formant frequencies (an acoustic cue to body size), in order to examine female looking responses to calls characterized by different combinations of these acoustic components (Charlton et al. 2008). The findings of this study suggested that the fundamental frequency of the male roar did not affect female perception of size-related formant information but also reinforced the idea that formants, being a long-term indicator of male body-size and hence genetic quality, are more important to red deer hinds as acoustic cues for mate assessment during the breeding season. Future mammal studies should follow this approach to examine the combined effect of variation in two separate components of a male mammals sexual display, using experimental setups in which the effects of competing male and female mating strategies, female hormonal state and female sampling strategies are controlled for.

In conclusion, female mate choice in mammals remains a complex research area requiring new and innovative experimental paradigms to elucidate what are often transient female preferences. Technological advances in digital sound production and playback equipment make female preferences for vocal characteristics, which can easily be isolated, a possible area for advancing our knowledge of female mate choice. Playback experiments can be used to present females with re-synthesized male calls, in which a vocal characteristic of interest is manipulated, to test their ability to perceive cues to male quality in mammal calls (Charlton et al. 2007a), and investigate whether they use these cues to assess male quality in a mate choice context (Charlton et al. 2007b.). Other technological advances now allow digital images to be realistically modified to test for visual preferences (Waitt et al. 2003) or to test the ability of females to link acoustic stimuli to visual representations of different male phenotypes (Ghazanfar et al. 2007). This experimental work should be followed up by fieldwork where mating behaviour can be observed and paternity analyses conducted under natural conditions to realistically determine the reproductive outcomes of male and female mating strategies. By combining knowledge gained from experimental and observational studies both in the field and laboratory we may ultimately hope to understand the link between mating preferences, female mate choice and reproductive success in mammals. Only a combination of carefully designed field and laboratory experiments will enable us to achieve this goal.

REFERENCES

Amos W, Wilmer JW, Kokko H (2001) Do female grey seals select genetically diverse mates? *Animal Behaviour* 62:157-164.

Andersson M (1994) Sexual Selection. Princeton, PRINCETON UNIVERSITY PRESS.

Apollonio M, Festabianchet M, Mari F, Riva M (1990) Site-Specific Asymmetries in Male Copulatory Success in a Fallow Deer Lek. *Animal Behaviour* 39:205-212.

Balmford A (1991) Mate Choice on Leks. *Trends in Ecology & Evolution* 6:87-&.

Balmford A, Albon S, Blakeman S (1992a) Correlates of Male Mating Success and Female Choice in a Lek- Breeding Antelope. *Behavioral Ecology* 3:112-123.

Balmford A, Rosser AM, Albon SD (1992b) Correlates of Female Choice in Resource-Defending Antelope. *Behavioral Ecology and Sociobiology* 31:107-114.

Beck CW, Powell LA (2000) Evolution of female mate choice based on male age: Are older males better mates? *Evolutionary Ecology Research* 2:107-118.

Borgia G (1979) Sexual selection and the evolution of mating system. In: *Sexual selection and reproductive competition* (Eds Blum M, Blum A), pp 19-80. New York: Academic Press.

Bradbury J (1977) Lek mating behavior in the hammer–headed bat. *Zeitschrift für Tierpsychologie* 45:225-255.

Brandt CA (1989) Mate Choice and Reproductive Success of Pikas. *Animal Behaviour* 37:118-132.

Bro-Jorgensen J (2002) Overt female mate competition and preference for central males in a lekking antelope. *Proceedings of the National Academy of Sciences of the United States of America* 99:9290-9293.

Brooks R (1998) The importance of mate copying and cultural inheritance of mating preferences. *Trends in Ecology & Evolution 13*:45-46.

Brown JL (1997) A theory of mate choice based on heterozygosity. *Behavioral Ecology* 8:60-65.

Brown JL, Eklund A (1994) Kin Recognition and the Major Histocompatibility Complex - an Integrative Review. *American Naturalist* 143:435-461.

Byers JA, Moodie JD, Hall N (1994) Pronghorn Females Choose Vigorous Mates. *Animal Behaviour* 47:33-43.

Candolin U (2003) The use of multiple cues in mate choice. *Biological Reviews* 78:575-595.

Carranza J (1995) Female Attraction by Males Versus Sites in Territorial Rutting Red Deer. *Animal Behaviour* 50:445-453.

Cassini MH, Fernandez-Juricic E (2003) Costs and benefits of joining South American sea lion breeding groups: testing the assumptions of a model of female breeding dispersion. *Canadian Journal of Zoology-Revue Canadienne De Zoologie* 81:1154-1160.

Chaine AS, Lyon BE (2008) Adaptive plasticity in female mate choice dampens sexual selection on male ornaments in the lark bunting. *Science* 319:459-462.

Charlton BD (2006) Female perception and use of size-related formant information in the roars of red deer stags (Cervus elaphus). PhD Thesis, University of Sussex.

Charlton BD, Reby D, McComb K (2007a) Female perception of size-related formant shifts in red deer, *Cervus elaphus*. *Animal Behaviour* 74:707-714.

Charlton BD, Reby D, McComb K (2007b) Female red deer prefer the roars of larger males. *Biology Letters* 3:382-385.

Charlton BD, Reby D, McComb K (2008) Effect of combined source (F0) and filter (formant) variation on red deer hind responses to male roars. *Journal of the Acoustical Society of America* 123 (5), 2936-2943.

Charlton BD, Reby D, McComb K (In submission to *Ethology*) Free-ranging red deer hinds use formant frequencies as acoustic cues to male body size and maturity.

Chism J, Rogers W (1997) Male competition, mating success and female choice in a seasonally breeding primate (Erythrocebus patas). *Ethology* 103:109-126.

Clarke FM, Faulkes CG (1999) Kin discrimination and female mate choice in the naked mole-rat Heterocephalus glaber. *Proceedings of the Royal Society of London Series B-Biological Sciences* 266:1995-2002.

Clutton-Brock TH, Albon S, Guinness FE (1982) Red Deer: Behavior and Ecology of Two Sexes. Chicago, UNIVERSITY OF CHICAGO PRESS.

Clutton-Brock TH, Hiraiwahasegawa M, Robertson A (1989) Mate Choice on Fallow Deer Leks. *Nature* 340:463-465.

Clutton-Brock TH, McComb K (1993) Experimental Tests of Copying and Mate Choice in Fallow Deer (Dama-Dama). *Behavioral Ecology* 4:191-193.

Clutton-Brock TH, Price OF, Maccoll ADC (1992) Mate Retention, Harassment, and the Evolution of Ungulate Leks. *Behavioral Ecology* 3:234-242.

Clutton-Brock TH, Vincent ACJ (1991) Sexual Selection and the Potential Reproductive Rates of Males and Females. *Nature* 351:58-60.

Conner DA, Whitworth MR (1985) The Ontogeny of Vocal Communication in the Pika. *Journal of Mammalogy* 66:756-763.

Cordero C, Eberhard WG (2003) Female choice of sexually antagonistic male adaptations: a critical review of some current research. *Journal of Evolutionary Biology* 16:1-6.

Cox CR, Leboeuf BJ (1977) Female Incitation of Male Competition - Mechanism in Sexual Selection. *American Naturalist* 111:317-335.

Craul M, Zimmermann E, Radespiel U (2004) First experimental evidence for female mate choice in a nocturnal primate. *Primates* 45:271-4.

Croll DA, Clark CW, Acevedo A, Tershy B, Flores S, Gedamke J, Urban J (2002) Bioacoustics: Only male fin whales sing loud songs - These mammals need to call long-distance when it comes to attracting females. *Nature* 417:809-809.

Darwin C (1871) The descent of man and selection in relation to sex. London, MURRAY.

Ditchkoff SS, Lochmiller RL, Masters RE, Hoofer SR, Van Den Bussche RA (2001) Major-histocompatibility-complex-associated variation in secondary sexual traits of white-tailed deer (Odocoileus virginianus): Evidence for good-genes advertisement. *Evolution* 55:616-625.

Drickamer LC, Gowaty PA, Holmes CM (2000) Free female mate choice in house mice affects reproductive success and offspring viability and performance. *Animal Behaviour* 59:371-378.

East ML, Burke T, Wilhelm K, Greig C, Hofer H (2003) Sexual conflicts in spotted hyenas: male and female mating tactics and their reproductive outcome with respect to age, social status and tenure. *Proceedings of the Royal Society of London Series B-Biological Sciences* 270:1247-1254.

East ML, Hofer H (1991) Loud Calling in a Female-Dominated Mammalian Society.2. Behavioral Contexts and Functions of Whooping of Spotted Hyaenas, Crocuta-Crocuta. *Animal Behaviour* 42:651-669.

Ehman KD, Scott ME (2002) Female mice mate preferentially with non-parasitized males. *Parasitology* 125:461-466.

Forrest TG, Raspet R (1994) Models of Female Choice in Acoustic Communication. *Behavioral Ecology* 5:293-303.

Fox EA (2002) Female tactics to reduce sexual harassment in the Sumatran orangutan (Pongo pygmaeus abelii). *Behavioral Ecology and Sociobiology* 52:93-101.

Gagneux P, Boesch C, Woodruff DS (1999) Female reproductive strategies, paternity and community structure in wild West African chimpanzees. *Animal Behaviour* 57:19-32.

Galimberti F, Boitani L, Marzetti I (2000) Female strategies of harassment reduction in southern elephant seals. *Ethology Ecology & Evolution* 12:367-388.

Gautier JP, Gautier A (1977) Communication in Old World primates. In: *How Animals Communicate* (Ed Sebeok TA), pp 890-964. Bloomington: Indiana University Press.

Gerald MS (2001) Primate colour predicts social status and aggressive outcome. *Animal Behaviour* 61:559-566.

Ghazanfar AA, Turesson HK, Maier JX, Van Dinther R, Patterson RD, Logothetis NK (2007) Vocal-Tract Resonances as Indexical Cues in Rhesus Monkeys. *Current Biology* 17:425-430.

Gould JL, Elliot SL, Masters CM, Mukerji J (1999) Female preferences in a fish genus without female mate choice. *Current Biology* 9:497-500.

Grafen A (1990) Biological Signals as Handicaps. *Journal of Theoretical Biology 144*:517-546.

Grahn M, Vonschantz T (1994) Fashion And Age In Pheasants - Age-Differences In Mate Choice. *Proceedings of the Royal Society of London Series B-Biological Sciences* 255:237-241.

Grob B, Knapp LA, Martin RD, Anzenberger G (1998) The major histocompatibility complex and mate choice: Inbreeding avoidance and selection of good genes. *Experimental and Clinical Immunogenetics* 15:119-129.

Halliday T (1983) The Study of Mate Choice. In: *Mate Choice* (Ed. By PPG Bateson), pp 3-32. Cambridge: Cambridge University Press.

Hamilton WD, Zuk M (1982) Heritable True Fitness and Bright Birds - a Role for Parasites. *Science* 218:384-387.

Hanggi EB, Schusterman RJ (1994) Underwater acoustic displays and individual variation in male harbour seals, *Phoca vitulina. Animal Behaviour* 48:1275-1283.

Hirotani A (1989) Social Relationships of Reindeer Rangifer-Tarandus During Rut - Implications for Female Choice. *Applied Animal Behaviour Science* 24:183-202.

Hogg JT (1987) Intrasexual Competition and Mate Choice in Rocky-Mountain Bighorn Sheep. *Ethology* 75:119-144.

Hohoff C, Franzen K, Sachser N (2003) Female choice in a promiscuous wild guinea pig, the yellow- toothed cavy (Galea musteloides). *Behavioral Ecology and Sociobiology* 53:341-349.

Horne TJ, Ylonen H (1996) Female bank voles (Clethrionomys glareolus) prefer dominant males; But what if there is no choice? *Behavioral Ecology and Sociobiology* 38:401-405.

Huck UW, Banks EM (1982) Differential Attraction of Females to Dominant Males - Olfactory Discrimination and Mating Preference in the Brown Lemming (Lemmus-Trimucronatus). *Behavioral Ecology and Sociobiology* 11:217-222.

Janson CH (1984) Female Choice and Mating System of the Brown Capuchin Monkey Cebus-Apella (Primates, Cebidae). *Zeitschrift Fur Tierpsychologie-Journal of Comparative Ethology* 65:177-200.

Jennions MD, Petrie M (1997) Variation in mate choice and mating preferences: A review of causes and consequences. *Biological Reviews of the Cambridge Philosophical Society* 72:283-327.

Johnston RE, Barot S (2004) Sexual selection, scent over-marking, and mate choice in golden hamsters. *Hormones and Behavior* 46:113-113.

Johnston RE, Sorokin ES, Ferkin MH (1997) Female voles discriminate males' over-marks and prefer top- scent males. *Animal Behaviour* 54:679-690.

Jordan WC, Bruford MW (1998) New perspectives on mate choice and the MHC. *Heredity* 81:239-245.

Kavaliers M, Colwell DD, Choleris E (1998) Parasitized female mice display reduced aversive responses to the odours of infected males. *Proceedings of the Royal Society of London Series B-Biological Sciences* 265:1111-1118.

Kavaliers M, Fudge MA, Colwell DD, Choleris E (2003) Aversive and avoidance responses of female mice to the odors of males infected with an ectoparasite and the effects of prior familiarity. *Behavioral Ecology and Sociobiology* 54:423-430.

Keane B (1990) The Effect of Relatedness on Reproductive Success and Mate Choice in the White-Footed Mouse, Peromyscus-Leucopus. *Animal Behaviour* 39:264-273.

Keddy AC (1986) Female Mate Choice in Vervet Monkeys (Cercopithecus-Aethiops-Sabaeus). *American Journal of Primatology* 10:125-134.

Keddy-Hector AC (1992) Mate Choice in Nonhuman-Primates. *American Zoologist* 32:62-70.

Kiley M (1972) The vocalisations of ungulates, their causation and function. *Zeitschrift für Tierpsychologie* 31:171-222.

Kirkpatrick M (1996) Good genes and direct selection in evolution of mating preferences. *Evolution* 50:2125-2140.

Kirkpatrick M, Ryan MJ (1991) The Evolution of Mating Preferences and the Paradox of the Lek. *Nature* 350:33-38.

Klappert K, Reinhold K (2003) Acoustic preference functions and sexual selection on the male calling song in the grasshopper Chorthippus biguttulus. *Animal Behaviour* 65:225-233.

Klinkova E, Hodges JK, Fuhrmann K, de Jong T, Heistermann M (2005) Male dominance rank, female mate choice and male mating and reproductive success in captive chimpanzees. *International Journal of Primatology* 26:357-384.

Kokko H, Brooks R, Jennions MD, Morley J (2003) The evolution of mate choice and mating biases. *Proceedings of the Royal Society of London Series B-Biological Sciences* 270:653-664.

Kokko H, Brooks R, McNamara JM, Houston AI (2002) The sexual selection continuum. *Proceedings of the Royal Society of London Series B-Biological Sciences* 269:1331-1340.

Kokko H, Lindstrom J (1996) Evolution of female preference for old mates. *Proceedings of the Royal Society of London Series B-Biological Sciences* 263:1533-1538.

Komers PE, Birgersson B, Ekvall K (1999) Timing of estrus in fallow deer is adjusted to the age of available mates. *American Naturalist* 153:431-436.

Lampe HM, Espmark YO (2003) Mate choice in Pied Flycatchers Ficedula hypoleuca: can females use song to find high-quality males and territories? *Ibis* 145:E24-E33.

Lisk RD, Huck UW, Gore AC, Armstrong MX (1989) Mate Choice, Mate Guarding and Other Mating Tactics in Golden- Hamsters Maintained under Seminatural Conditions. *Behaviour* 109:58-75.

Lynch KS, Rand AS, Ryan MJ, Wilczynski W (2005) Plasticity in female mate choice associated with changing reproductive states. *Animal Behaviour* 69:689-699.

Maestripieri D, Roney JR (2005) Primate copulation calls and postcopulatory female choice. *Behavioral Ecology* 16:106-113.

Manson JH (1995) Do Female Rhesus Macaques Choose Novel Males. *American Journal of Primatology* 37:285-296.

Matsumoto-Oda A (1999) Female choice in the opportunistic mating of wild chimpanzees (Pan troglodytes schweinfurthii) at Mahale. *Behavioral Ecology and Sociobiology* 46:258-266.

Maynard Smith J (1991) Theories of Sexual Selection. *Trends in Ecology & Evolution* 6:146-151.

Mays HL, Hill GE (2004) Choosing mates: good genes versus genes that are a good fit. *Trends in Ecology & Evolution* 19:554-559.

McComb KE (1991) Female Choice for High Roaring Rates in Red Deer, *Cervus elaphus*. *Animal Behaviour* 41:79-88.

McElligott AG, O'Neill KP, Hayden TJ (1999) Cumulative long-term investment in vocalization and mating success of fallow bucks, Dama dama. *Animal Behaviour* 57:1159-1167.

Meagher S, Penn DJ, Potts WK (2000) Male-male competition magnifies inbreeding depression in wild house mice. *Proceedings of the National Academy of Sciences of the United States of America* 97:3324-3329.

Mech SG, Dunlap AS, Wolff JO (2003) Female prairie voles do not choose males based on their frequency of scent marking. *Behavioural Processes* 61:101-108.

Mitani JC (1985) Sexual Selection and Adult Male Orangutan Long Calls. *Animal Behaviour* 33:272-283.

Miura S (1984) Social-Behavior and Territoriality in Male Sika Deer (Cervus- Nippon Temminck 1838) During the Rut. *Zeitschrift Fur Tierpsychologie-Journal of Comparative Ethology* 64:33-73.

Mobley JR (1988) Responses of wintering humpback whales (*Megaptera novaeangliae*) to playback recordings of winter and summer vocalizations and of synthetic sound. *Behavioral Ecology and Sociobiology* 23:211-233.

Moller AP, Jennions MD (2001) How important are direct fitness benefits of sexual selection? *Naturwissenschaften* 88:401-415.

Moore PJ, Moore AJ (2001) Reproductive aging and mating: The ticking of the biological clock in female cockroaches. *Proceedings of the National Academy of Sciences of the United States of America* 98:9171-9176.

Mountjoy DJ, Lemon RE (1996) Female choice for complex song in the European starling: A field experiment. *Behavioral Ecology and Sociobiology* 38:65-71.

Murphy CG, Gerhardt HC (2000) Mating preference functions of individual female barking treefrogs, Hyla gratiosa, for two properties of male advertisement calls. *Evolution* 54:660-669.

Nefdt RJC, Thirgood SJ (1997) Lekking, resource defense, and harassment in two subspecies of lechwe antelope. *Behavioral Ecology* 8:1-9.

Neff BD, Pitcher TE (2005) Genetic quality and sexual selection: an integrated framework for good genes and compatible genes. *Molecular Ecology* 14:19-38.

Oconnell SM, Cowlishaw G (1994) Infanticide Avoidance, Sperm Competition and Mate Choice - the Function of Copulation Calls in Female Baboons. *Animal Behaviour* 48:687-694.

Page B, Goldsworthy SD, Hindell MA, McKenzie J (2002) Interspecific differences in male vocalizations of three sympatric fur seals (Arctocephalus spp.). *Journal of Zoology* 258:49-56.

Palombit RA, Seyfarth RM, Cheney DL (1997) The adaptive value of 'friendships' to female baboons: experimental and observational evidence. *Animal Behaviour* 54:599-614.

Parga J (2002) Male dominance rank reversals during the breeding season in ringtailed lemurs (Lemur catta): changes resulting from female mate choice. *American Journal of Physical Anthropology*:123-123.

Paterson S, Pemberton JM (1997) No evidence for major histocompatibility complex-dependent mating patterns in a free-living ruminant population. *Proceedings of the Royal Society of London Series B-Biological Sciences* 264:1813-1819.

Paterson S, Wilson K, Pemberton JM (1998) Major histocompatibility complex variation associated with juvenile survival and parasite resistance in a large unmanaged ungulate population (Ovis aries L.). *Proceedings of the National Academy of Sciences of the United States of America* 95:3714-3719.

Paul A (2002) Sexual selection and mate choice. *International Journal of Primatology* 23:877-904.

Penn DJ (2002) The scent of genetic compatibility: Sexual selection and the major histocompatibility complex. *Ethology* 108:1-21.

Penn DJ, Potts WK (1999) The evolution of mating preferences and major histocompatibility complex genes. *American Naturalist* 153:145-164.

Pereira ME, Weiss ML (1991) Female Mate Choice, Male Migration, and the Threat of Infanticide in Ringtailed Lemurs. *Behavioral Ecology and Sociobiology* 28:141-152.

Pillay N (2000) Female mate preference and reproductive isolation in populations of the striped mouse Rhabdomys pumilio. *Behaviour* 137:1431-1441.

Poole JH (1989) Mate Guarding, Reproductive Success and Female Choice in African Elephants. *Animal Behaviour* 37:842-849.

Poole KG, Murphy CG (2007) Preferences of female barking treefrogs, Hyla gratiosa, for larger males: univariate and composite tests. *Animal Behaviour* 71:513-524.

Pope TR (1990) The Reproductive Consequences of Male Cooperation in the Red Howler Monkey - Paternity Exclusion in Multimale and Single- Male Troops Using Genetic-Markers. *Behavioral Ecology and Sociobiology* 27:439-446.

Potts WK, Manning CJ, Wakeland EK (1991) Mating patterns in seminatural populations of mice influenced by MHC genotype. *Nature* 352:619-621.

Preston BT, Stevenson IR, Wilson K (2003) Soay rams target reproductive activity towards promiscuous females' optimal insemination period. *Proceedings of the Royal Society of London Series B-Biological Sciences* 270:2073-2078.

Price T, Schluter D, Heckman NE (1993) Sexual Selection When the Female Directly Benefits. *Biological Journal of the Linnean Society* 48:187-211.

Qvarnstrom A (2001) Context-dependent genetic benefits from mate choice. *Trends in Ecology & Evolution* 16:5-7.

Radespiel U, Dal Secco V, Drogemuller C, Braune P, Labes E, Zimmermann E (2002) Sexual selection, multiple mating and paternity in grey mouse lemurs, Microcebus murinus. *Animal Behaviour* 63:259-268.

Radespiel U, Zimmermann E (2003) The influence of familiarity, age, experience and female mate choice on pregnancies in captive grey mouse lemurs. *Behaviour* 140:301-318.

Real L (1990) Search Theory and Mate Choice.1. Models of Single-Sex Discrimination. *American Naturalist* 136:376-405.

Reale D, Bousses P, Chapuis JL (1996) Female-biased mortality induced by male sexual harassment in a feral sheep population. *Canadian Journal of Zoology-Revue Canadienne De Zoologie* 74:1812-1818.

Reby D, McComb K (2003) Anatomical constraints generate honesty: acoustic cues to age and weight in the roars of red deer stags. *Animal Behaviour* 65:519-530.

Reeceengel C (1988) Female Choice of Resident Male Rabbits Oryctolagus-Cuniculus. *Animal Behaviour* 36:1241-1242.

Reynolds JD, Gross MR (1990) Costs and Benefits of Female Mate Choice - Is There a Lek Paradox. *American Naturalist* 136:230-243.

Rolland C, MacDonald DW, De Fraipont M, Berdoy M (2003) Free female choice in house mice: Leaving best for last. *Behaviour* 140:1371-1388.

Rulicke T, Chapuisat M, Homberger FR, Macas E, Wedekind C (1998) MHC-genotype of progeny influenced by parental infection. *Proceedings of the Royal Society of London Series B-Biological Sciences* 265:711-716.

Scheuber H, Jacot A, Brinkhof MWG (2004) Female preference for multiple condition-dependant components of a sexually selected signal. *Proceedings Of The Royal Society B-Biological Sciences* 271:2453-2457.

Searcy WA, Andersson M (1986) Sexual Selection and the Evolution of Song. *Annual Review of Ecology and Systematics* 17:507-533.

Semple S (1998) The function of Barbary macaque copulation calls. *Proceedings of the Royal Society of London Series B-Biological Sciences* 265:287-291.

Setchell JM, Dixson AF (2001) Arrested development of secondary sexual adornments in subordinate adult male mandrills (Mandrillus sphinx). *American Journal of Physical Anthropology* 115:245-252.

Shapiro LE, Dewsbury DA (1986) Male-Dominance, Female Choice and Male Copulatory-Behavior in 2 Species of Voles (Microtus-Ochrogaster and Microtus-Montanus). *Behavioral Ecology and Sociobiology* 18:267-274.

Simmons LW, Zuk M (1992) Variability In Call Structure And Pairing Success Of Male Field Crickets, Gryllus-Bimaculatus - The Effects Of Age, Size And Parasite Load. *Animal Behaviour* 44:1145-1152.

Smuts BB (1987) Sexual competition and mate choice. In: *Primate Societies* (Eds Smuts BB, Cheney DL, Seyfarth RM, Wrangham RW, Struhsaker TT), pp 385-399. Chicago: University of Chicago Press.

Soltis J (1999) Measuring male-female relationships during the mating season in wild Japanese macaques (Macaca fuscata yakui). *Primates* 40:453-467.

Soltis J, Mitsunaga F, Shimizu K, Nozaki M, Yanagihara Y, DomingoRoura X, Takenaka O (1997) Sexual selection in Japanese macaques.2. female mate choice and male-male competition. *Animal Behaviour* 54:737-746.

Stumpf RM, Boesch C (2005) Does promiscuous mating preclude female choice? Female sexual strategies in chimpanzees (Pan troglodytes verus) of the Tai National Park, Cote d'Ivoire. *Behavioral Ecology and Sociobiology* 57:511-524.

Takahata Y, Huffman MA, Suzuki S, Koyama N, Yamagiwa J (1999) Why dominants do not consistently attain high mating and reproductive success: A review of longitudinal Japanese macaque studies. *Primates* 40:143-158.

Thomas SA (2002) Scent marking and mate choice in the prairie vole, Microtus ochrogaster. *Animal Behaviour* 63:1121-1127.

Thornhill R (1976) Sexual Selection and Nuptial Feeding-Behavior in Bittacus- Apicalis (Insecta-Mecoptera). *American Naturalist* 110:529-548.

Todt D, Hammerschmidt K, Ansorge V, Fischer J (1995) The vocal behaviour of Barbary macaques (Macaca sylvanus): Call features and their performance in infants and adults. In: *Current Topics in Primate Vocal Communication* (Eds Zimmermann E, Newman JD, Jurgens U, pp 141-160. New York: Plenum Press.

Tomihara K (2005) Selective approach to a male and subsequent receptivity to mounting comprise mate-choice behavior of female mice. *Japanese Psychological Research* 47:22-30.

Trivers RL (1972) Parental investment and sexual selection. In: *Sexual selection and the descent of man* (Ed Campbell B), pp 136-179. Chicago: Aldine Press.

Tyack P (1981) Interactions between Singing Hawaiian Humpback Whales and Conspecifics Nearby. *Behavioral Ecology and Sociobiology* 8:105-116.

Tyack P (1983) Differential response of humpback whales, *Megaptera novaeangliae,* to playback of song or social sounds. *Behavioral Ecology and Sociobiology* 13:49-55.

Valsecchi P, Razzoli M, Choleris E (2002) Influence of kinship and familiarity on the social and reproductive behaviour of female Mongolian gerbils. *Ethology Ecology & Evolution* 14:239-253.

van Doorn GS, Weissing FJ (2004) The evolution of female preferences for multiple indicators of quality. *American Naturalist* 164:173-186.

Veen T, Borge T, Griffith SC, Saetre GP, Bures S, Gustafsson L, Sheldon BC (2001) Hybridization and adaptive mate choice in flycatchers. *Nature* 411:45-50.

Vorobieva E, Hodges K, Heistermann M (2004) The relative importance of male dominance rank and female mate choice on male mating and reproductive success in male chimpanzees. *Hormones and Behavior* 46:124-124.

Wagner WE (1998) Measuring female mating preferences. *Animal Behaviour* 55:1029-1042.

Waitt C, Little AC, Wolfensohn S, Honess P, Brown AP, Buchanan-Smith HM, Perrett DI (2003) Evidence from rhesus macaques suggests that male coloration plays a role in female primate mate choice. *Proceedings of the Royal Society of London Series B-Biological Sciences* 270:S144-S146.

Welch AM (2003) Genetic benefits of a female mating preference in gray tree frogs are context-dependent. *Evolution* 57:883-893.

Welker C, Hohmann H, Schaferwitt C (1990) Significance of Kin Relations and Individual Preferences in the Social-Behavior of Cebus-Apella. *Folia Primatologica* 54:166-170.

West PM, Packer C (2002) Sexual selection, temperature, and the lion's mane. *Science* 297:1339-1343.

Wiley RH, Poston J (1996) Perspective: Indirect mate choice, competition for mates, and coevolution of the sexes. *Evolution* 50:1371-1381.

Willis C, Poulin R (2000) Preference of female rats for the odours of non-parasitised males: the smell of good genes? *Folia Parasitologica* 47:6-10.

Wolff JO (1998) Breeding strategies, mate choice, and reproductive success in American bison. *Oikos* 83:529-544.

Wolff JO, Macdonald DW (2004) Promiscuous females protect their offspring. *Trends in Ecology & Evolution* 19:127-134.

Yamazaki K, Beauchamp GK, Curran M, Bard J, Boyse EA (2000) Parent-progeny recognition as a function of MHC odortype identity. *Proceedings of the National Academy of Sciences of the United States of America* 97:10500-10502.

Zahavi A (1975) Mate Selection - Selection for a Handicap. *Journal of Theoretical Biology* 53:205-214.

Zeh JA, Zeh DW (1996) The evolution of polyandry I: Intragenomic conflict and genetic incompatibility. *Proceedings of the Royal Society of London Series B-Biological Sciences* 263:1711-1717.

Zeh JA, Zeh DW (1997) The evolution of polyandry II: Post-copulatory defenses against genetic incompatibility. *Proceedings Of The Royal Society B-Biological Sciences* 264:69-75.

In: Animal Behavior: New Research
Editors: E. A. Weber, L. H. Krause

ISBN: 978-1-60456-782-3
© 2008 Nova Science Publishers, Inc.

Chapter 3

PARTITION TEST AND SEXUAL MOTIVATION IN MALE MICE

N. N. Kudryavtseva[*]

Neurogenetics of Social Behavior Sector, Institute of Cytology and Genetics, Siberian Department of Russian Academy of Sciences, Novosibirsk, 630090, Russia

ABSTRACT

Theoretical analysis of own and literature investigations of sexual motivation with the use of the partition test [Kudryavtseva, 1987, 1994] in male mice was carried out. It has been shown that appearance of a receptive female in the neighboring compartment of common cage separated by perforated transparent partition produces the enhancement of testosterone level in blood and stimulates the behavioral activity near partition as a reaction to the receptive female in naive males. In many studies this behavioral activity is considered as sexual motivation, arising in this experimental context in male mice. The lack of correlation between behavioral parameters and gonad reaction of males on receptive female, uninterconnected changes of these two parameters as well as the lack of sexual behavior between naive male and female when partition is removed cast doubt on this data interpretation. It has been supposed that in naive males behavioral reaction to a receptive female is induced by positive incentive – odor of the female associated with nursing and warmth from mother and other females which look after posterity. Short-term increase of the level of testosterone (possessing rewarding properties) is innate stimulus-response reaction which stimulates and prolongs behavioral interest of male to receptive female. It has been supposed that after sexual experience female odor is associated in experienced males with sexual behavior directed to the sexual partner and resulted in the formation of sexual motivation. The data are considered also in the light of the theory of motivated behavior including "liking", "wanting" and "learning" [Robinson and Berridge, 1993, 2000].

---

[*] natnik@bionet.nsc.ru

Keywords: partition test; sexual motivation; sexual arousal; mice

INTRODUCTION

Since the 1950s influence of pheromones has been investigated in mice with the use of a small cage divided into two compartments by a partition (wire mesh) allowing the animals to see, hear and smell each other but preventing physical contact. It has been shown that a strange male located in the neighboring compartment can block incipient pregnancy of the female without direct physical contact [Bronson and Eleftheriou, 1969; Bruce, 1960] by inducing changes in the hormonal background. In turn, female pheromones induce increase plasma testosterone and other hormones in male distantly [6, 7, 45], the intensity of enhancement is different in male mice of various strains [Amstislavskaya et al., 1990; Naumenko et al., 1983]. Neuroendocrine and neurochemical mechanisms of male sexual arousal induced by the presence of receptive female have been investigated [Naumenko et al., 1991; Naumenko et al., 1992; Naumenko et al., 1983; Osadchuk and Naumenko, 1981]. Distant influence of males of different social status on the hormonal background of each other [Bronson et al., 1969; Bronson et al., 1973] as well as behavioral reaction of males and females on each other depending on the stage of female estrus cycle [Steel, 1983] were also studied. The similar cage with perforated transparent partition was used to form aggressive and submissive types of behavior in male mice in daily intermale confrontations [Kudryavtseva, 1991]. As a result, the "partition test" [Kudryavtseva, 1987; Kudryavtseva, 1994] was suggested as a tool for estimating behavioural reactivity of mice to the conspecific behind the partition dividing the experimental cage into equal parts. Number of approaches to the partition and total time spent near the partition, when mice touched the partition with their fore parts (nose, paws) were scored as indices of reacting to the partner in the neighboring compartment. The average time spent near the partition during one approach was evaluated by the ratio of the total time to the number of approaches to or touches of the partition. The behavioral response of animals to conspecific was shown to differ depending on physiological and psychological status of an individual, his social experience, the type of partner in the neighboring compartment and strain. Many experiments provided support in evidence of the fact that the "partition" test could be useful in experiments designed to study the mechanisms of sociability, anxiety, olfaction, aggressive and sexual behavior, cognition, psychoemotional disorders as well as in pharmacological studies of the above [Kudryavtseva, 2003].

It became obvious that behavioral response to conspecific appears to reflect the motivation component of many behaviors based on previous experience or arising as instinctive response to innate specific (unconditional) stimulus (in this case pheromones) or partner behavior. The proof was obtained for anxiety and aggression. In male mice, behavioral parameters in the partition test were shown to correlate significantly with parameters in the elevated plus-maze test [Kudryavtseva et al., 2002], which is sensitive to the action of anxiolytics [Rodgers and Cole, 1994]. It was shown that the less time a male spent responding to the partner in the neighboring compartment, the less time it spent in the open arms of the plus maze. This fact indicated that the partition test could be used to estimate level of anxiety in relevant experimental context. The same was found in the study of aggressive motivation: the more time a male spent near the partition responding to the other male behind it, the higher level of aggression this male exhibited in the subsequent encounter, when the partition was removed and animals could freely contact each other [Kudryavtseva, 1989; Kudryavtseva et al., 2000]. In this case, as in other classical models of motivated behavior [Menning, 1982], the level of motivation was evaluated by the intensity of the corresponding consummatory act arising from the demonstration of intention (for example, aggressive behavior) or the absence of such (for example, in the case of anxiety and fear). It is these correlations that allowed using the partition test for the evaluation of the level of motivation in animals.

SEXUAL MOTIVATION

It was shown in many studies that a receptive female in the neighboring compartment of a common cage enhances the level of testosterone in sexually naïve male [Naumenko et al., 1983; Osadchuk and Naumenko, 1981]. This fact served as a base for the assumption that the test could be used for the measurement of sexual motivation. The first data proved that behavioral activity of males near the partition at presence of a receptive female was significantly higher than of a male [Bakshtanovskaya and Khodakova 1992; Kudryavtseva et al., 1997]. However it was known that receptive female is not the only driver of blood testosterone in males. Stress was shown to be accompanied by an increase in testosterone level in males too [Heiblum et al., 2000]. Notably, testosterone level increases proactively before aggressive confrontations [Suay et al., 1999].

As early as in the first experiments in male mice with different social status some doubts appeared that it is possible to interpret the partition behavior of males in the presence of receptive female in the context of sexual motivation. For example, aggressive males with the experience of victories in daily agonistic interactions were shown practically not to increase behavioral activity near the partition as a reaction to a female, which was put into the neighboring compartment immediately after the other male [Bakshtanovskaya and Khodakova 1992; Kudryavtseva et al., 1997]. At the same time, control and defeated animals doubtlessly displayed an interest in a receptive female. Preliminary data pointed to the absence of testosterone level increase in aggressive animals in response to the exhibition of a receptive female in contrast to the other two groups. This phenomenon was later explained by an inhibiting effect of chronic experience of aggression (forming an enhanced level of aggressive motivation) on sexual motivation or by an impaired sexual arousal in aggressive males. However the fact of the impairment of sexual motivation in aggressive males was paradoxical and contradicted extensive literature data indicating that it is the enhanced testosterone level that underlies the demonstration of aggression and reproductive success of dominant males [Bahrke et al., 1996, Mazur and Booth, 1998; Rubinow and Schmidt, 1996]. But the second experiment on another mouse strain provided the verification of earlier data [Kudryavtseva et al., 2004] on sexual motivation deficit in aggressive male mice.

Subsequently, however, after additional experiments the above results were interpreted not as impairment of sexual arousal *per se* but as impaired social recognition in aggressive males at social "male-female" and "male-male" interactions [Bondar' and Kudryavtseva, 2005]. In experiment [Amikishieva and Ovsiukova, 2004], with prior release from aggressive motivation, when the aggressive male was put in an individual cage for a day taken away from social confrontations, like control, aggressive males were found to display adequate behavioral response to the presence of a receptive female. Still, sexual motivation estimated by partition test in such males on the 30th minute of measurement was reduced as compared with control, in which behavioral response to a female remained practically unchanged [Amikishieva and Ovsiukova, 2004].

Obviously, to speak with certainty about changes in sexual motivation in each particular case it is necessary to obtain solid proof and conduct additional experiments. For example, it is necessary to find correlations between parameters of behavior near the partition in response to a receptive female prior to sexual interactions with certain forms of sexual behavior (sniffing, mountings, copulations, intromissions etc.) in the following direct contact with the sexual partner. Or a correlation is to be found between the parameters of behavior near the partition and some parameters of the gonadal hormone status, for instance, testosterone level. This would be a direct indication that the partition test could be used for measurement of the level of masculine sexual motivation in a particular experimental context.

However, the near-partition behavior in response toward a receptive female and hormone reaction turned out to be uncorrelated parameters. Moreover, with the partition removal sexual interaction between males and females failed to develop, at least, during the time of observation. A single study [Amikishieva and Ovsiukova, 2004] finally detected a correlation between the parameters of male behavior near the partition and some behavioral patterns towards the female (mostly sniffing) after the removal of the partition. This study also failed to demonstrate pronounced sexual behavior. It is not clear whether naïve or sexually experienced males were used in this study.

Analysis of literature data of the authors [Amstislavskaya, 2000; Amstislavskaya T.G., Popova, 2004; Popova et al., 1998; Kudryavtseva, 2003] who supposed that they were measuring sexual motivation by the partition test discriminating two components, behavioral and hormonal, also produced some doubtful moments:

1. In the above studies no significant correlations were found between behavioral parameters at the partition in naïve males in response to a receptive female and blood testosterone level [Amstislavskaya, 2000, Amstislavskaya and Khrapova, 2002];

2. Testosterone level increased to a maximum 20 minute after the introduction of the receptive female in the neighboring compartment and then (over half an hour) decreased to a control level. At the same time, behavioral reaction started immediately and remained at a high level after an hour [Amstislavskaya, 2000, Popova et al., 1998]. To some extent, this points to a lack of correlation between behavioral and hormonal components and to independent dynamics of these two characteristics in males;

3. In male mice of some strains a lack of both behavioral and hormonal reactions was found after 30 minutes exposure to BALB/c receptive female [Amstislavskaya and Khrapova, 2002]. Earlier data obtained with the use of this experimental approach indicated that 40 minutes after the exhibition of a female in the neighboring compartment males not all strains reacted by a rise in testosterone level [Naumenko et al., 1983];

4. In male rats, the behavioral activity was enhanced pronouncedly with practically no change in testosterone level in response to a female put into the neighboring compartment [Bulygina et al., 2001];

5. Under effect of some serotonin preparations one component of sexual motivation (hormonal or behavioral) changed and the other remained unchanged, for example, pharmacological stimulation of 5-HT1B receptors reduced the time of behavioral response to a female and had no effect on testosterone level [Amstislavskaya, 2006; Rodgers and Cole? 1994];

6. Exposure to stress could also change one component of sexual motivation and produce no effect on the other. For example, restriction for five hours that took place a day before the test did not change the intensity of behavioral response to receptive female but significantly diminished testosterone level [Amstislavskaya, 2006].

The reported data indicated that behavioral and hormonal components of the response to the presence of a receptive female in naïve male might act as independent variables (rather than an indivisible reaction), which is likely to change according to its own laws. The direct indication that the term "sexual motivation" in the given experimental context bears a great amount of hypothesizing was the lack of demonstration of sexual behavior by naïve males after the removal of the partition, when males and females could freely contact each other.

HYPOTHESIS

The effort to interpret the observations and answer the question "What is that the partition test actually measures in sexually naïve males?" and also to offer, with regard for own and literature data, a logical explanation of what is taking place in the cage at distant interaction of naïve males and receptive females has lead to a quite unexpected conclusion. In given experimental context, the partition test is most likely not to measure sexual motivation. To explain the absence of sexual interaction between mice of different sexes one is to assume that sexually naïve males probably react to the odor that was positively rewarded in the babyhood – nursing and warmth provided by the mother and other females who looked after the progeny together. Males like mother's odor (sort of "Oedipus situation"), they spend a lot of time at the partition trying to get over with the reaction to the female lasting more than an hour. This is evidence of the appeal of female odor for males and positive emotions of the males. It is then reasonable to assume that for mouse strains whose males withheld reaction to receptive female of a other strain [Amstislavskaya and Khrapova, 2002] the female odor was alien and too dissimilar to mother's odor so no behavioral reaction was exhibited. This also explains the absence of a correlation with testosterone level and the absence of sexual behavior.

However, if one agrees that in naïve males exposed to the a receptive female the partition test does not measure sexual motivation, it has to be accepted that an increase of testosterone is independent of the behavior, arises instinctively as innate response to a species-specific incentive – the odor of sexual partner. It was shown experimentally that in naïve males a behavioral reaction to the female and testosterone increase could be unrelated parameters. Alternatively, a short-term rise of testosterone level in response to a receptive female could be explained by a non-specific response to positive emotion, in which testosterone may be involved. Support for this premise could be found in the literature that shows the rewarding effects of testosterone [Alexander et al., 1994; Clark et al., 1996; Wood et al., 2004]. However it was natural to assume that the behavioral and hormonal components of the response to a receptive female may become interconnected upon acquisition of sexual experience.

POSSIBLE SCENARIO OF EVENTS
IN THE GIVEN EXPERIMENTAL CONTEXT

In naïve males, behavioral reaction *(sexual activation)* to the presence of a female arises immediately when the female is placed in the neighboring compartment of the cage. The reaction is induced by an incentive positively reinforced in the babyhood (odor of mother and other females). Additional component of the response could be the demonstration of communicative and exploratory activities toward a strange partner.

An increase in testosterone *(sexual arousal)* occurs in 20-40 min [Amstislavskaya, 2006; Naumenko et al., 1983; Popova et al., 1998], which can be regarded as a delayed response to the female odor. This response is instinctive, based on stimulus-response behavior, where the stimulus—the female odor—is innate and species-specific. Then over half an hour the testosterone level decreases but the behavioral reaction lasts for a long time. A question arises: why so long? In the experiments focusing on the duration of the host male response to a strange male exhibited in the neighboring compartment of common cage behavioral activity at the partition as a response to the male decreased relatively quickly [Kudryavtseva, 1987]. The prolonged behavioral reaction to the female is likely to be rewarded by the enhanced level of testosterone, which is capable of inducing positive emotional reaction all by itself. It may be assumed that even a short rise of testosterone level tints a distant contact with a female with positive emotions, creating positive reinforcement even distantly.

With maturity accompanied by an increase of testosterone level at the encounter with female odor as a source of sexual arousal there arise associative relations that form goal-directed behavior of female search aimed at obtaining positive emotions from enhanced blood testosterone level. Prolonged sexual excitation without coping could lead to the development of a sexual pathology.

Once the first sexual experience occurred, the male is aware of the positive implications of sexual catharsis that brings pleasure and relief over satisfaction. These positive implications build up *sexual motivation*, which is conceived as a behavior directed to satisfying the existing desire. Thus, motivation, which forms the base of goal-directed behavior, comes to existence only with experience [Simonov, 1987]. The use of the term "sexual motivation" for the description of naïve male behavior in response to a female appears to be somewhat misguiding in the frame of this theory as naïve males are lacking sexual experience.

Preliminary investigation of experienced and naïve males in the experiment under consideration showed that the overwhelming majority of experienced males (who participated in reproduc-tion and had sexual successes) after a 5-days refractory period at room lighting from the beginning demonstrated highly pronounced behavioral activity at the partition in response to a receptive female and then a expressed sexual behavior (anogenital sniffing, mountings and copulations) during 20 minutes of observation from the moment of partition removal. In naïve males (who demonstrate weaker reaction to the female) after partition removal aggressive attacks on the female were observed as well as considerably reduced sniffing activity compared with experienced males. Some of the naïve males made mounting attempts but after 10 minutes both males and females lost interest in each other. In general, it can be stated that naïve males do have interest in the female while lacking sexual motivation (as judged by the missing consum-matory act) or having such at a significantly lower level than sexually experienced males. At the beginning it was difficult even to decide whether it was sexual interest to the female or just interest in a new partner, in which anogenital sniffing is also present. The main conclusion made on the basis of these observations was the necessity to use a sexually experienced female as an incentive rather than a naïve female since the behavior of the latter, who is unaware of what the naïve male is wanting from her, may produce a inhibiting effect on the demonstration of his subsequent sexual behavior in naive male mice.

LIKING, WANTING, LEARNING

Motivation theories advanced in different years were reviewed extensively by K.S. Berridge [Berridge, 2004]. Jointly with T.E. Robinson he suggested a biopsychological theory of addictions, which was further elaborated in later years [Robinson and Berridge, 1993, 2000]. In the study of psychological and neurobiological prerequisites of drug addictions the authors separated two motivation components: "liking" (getting pleasure from drugs) and "wanting" (drug abuse, motivation to take the drugs). It was suggested that the two components may have different neural regulation. "Liking" is regulated by the brain's opioidergic and GABA-benzodiazepine systems while the regulation of "wanting " involves dopaminergic mediation paths with different brain structures being involved in the processes. Both "wanting" and "liking" may operate outside of subject's conscious awareness [Berridge, 1996]. As the drug addiction develops (with experience acquired - learning) the balance between the two components may shift to "wanting" [Robinson and Berridge, 1993, 2000].

This conception was extended to food motivation – food consumption is viewed as positive reinforcement, in which two independent components can also be separated: "liking" – pleasure from the food (hedonistic component) and "wanting" – proper food motivation [Berridge, 1996; Finlayson et al., 2007; Mela, 2006], which are supposed to have different mechanisms. For instance, we are ready to eat the food that we do not like when we are hungry. Or we are not hungry but will eat the food because we like it.

The components of motivational behavior "liking", "wanting" and "learning" appear to be applicable for the explanation of the mechanisms of sexual motivation. The role of learning (or experience) in sexual behavior has been studied substantially [Pfaus et al., 2001]. Sexual behavior is commonly understood as the behavior aimed at getting the pleasure from copulation, the notion of which individuals gain by different means from the initial experience. Learning (sexual experience) is a key to all further forms of sexual behavior such as search, preference for certain partner, obstacle overcoming, courtship, sexual arousal conditioned by perceived positive reinforcement, and copulation [Pfaus et al., 2001]. It may be assumed that repeated sexual experience is forming the motivation directed to both getting pleasure ("liking") and satisfying wanting in case of prolonged refractory period.

In sexually naïve males the behavioral reaction to a receptive female involves "liking" and at first is not associated with sexual behavior. "Liking" is driven by female odor and positive emotions from enhanced testosterone level in her presence. Sexual experience accompanied by powerful positive reinforcement on the level of physiological reactions resulted from sexual interactions leads to the formation of motivation for its repeated occurrence, which turns into the need (wanting). B. Everitt [Everitt. 1990] has presented evidence of different mechanisms of sexual arousal and its realization (intromissions and ejaculation).

It may be assumed that the two components of sexual motivated behavior, "liking" and "wanting", could manifest themselves in association as they are associated forms of the same state, but at the same time they can exist all by themselves, independently of each other. For example, in naïve males there is only "liking". In sexually experienced males there are both "wanting" and "liking", but "liking" implicates sexual contact. It may be anticipated that these males are most likely to show a correlation between behavioral activity at the partition and testosterone level. It is expected that correlations with sexual behavior should be found in males with the matured sexual drive and a sufficiently long refractory period after the last copulation.

Studies of male behavior in response to a female depending on the presence or absence of sexual experience of the male produced implicit evidence in support of this hypothesis. Naïve male rats having to make a choice were shown to prefer an incentive from the female but this preference was reduced as compared with what was demonstrated by experienced males. A similar result was obtained in the test when a male could see, hear and smell an experienced male or a receptive female without contacting them physically [53]. It was shown that it took sexually experienced males less time to run up to a receptive or nonreceptive female [Lo´pez et al., 1999] and to make a correct choice of an estrus female in the T-maze as compared with naïve males [Mazur A., Booth 1998 40]. Some authors considered the behavior of a naïve male toward a receptive female as unconditioned sexual incentive motivation, while the reaction of a sexually experienced male – as conditioned sexual incentive motivation [Agmo, 2003; Pfaus et al., 2001].

THE FEATURES OF THE PARTITION TEST APPLICATION IN THE STUDY OF SEXUAL MOTIVATION IN MALE MICE

Adequate Partner Choice

In the studies of sexual motivation in mice of different strains, strange as it may seem, it is most difficult to choose an adequate sexual partner – the female. Such studies [Amstislavskaya and Khrapova, 2002 Naumenko et al., 1983] commonly used a receptive female of BALB/c strain, which was supposed to have an attractive and powerful odor for male mice. However, males of not all strains reacted to BALB/c female by the enhancement of testosterone level and behavioral activity in the partition test. The obvious conclusion is that sexual motivation in these males is diminished. But it is well known that mice of inbred strains are distinct as regards many psychophysiological characteristics: emotionality, locomotor and exploratory activities, anxiety, sensitivity to olfactory incentives and ability to discriminate such incentives etc. Any of these features may produce an effect on the response intensity under exposure to a strange receptive female placed in the neighboring compartment of the cage. For instance, it may be assumed that some males do and other males do not feel the smell, which may have nothing to do with the mechanisms of sexual motivation as such. Back in 1988 S. N. Novikov in his excellent monograph "Pheromones and reproduction of mammals" [Novikov. 1988] analyzed data on testosterone level increase in male mice of different strains in the presence of a receptive female [Naumenko rt al., 1983] and raised the questions, the convincing answers to which have not been found yet: "What is the relative role of olfaction in the regulation of this effect in different strains?", "Is the female genotype of importance for the manifestation of sexual activation effect?" and "What neurophysiological mechanisms underlie low reactivity of males of some strains?" It is doubtful that males of the strains that did not react to the BALB/c females breed worse than males of other strains or that sexual motivation of such males is reduced. This example illustrates the complexity of the task of sexual partner selection for the experiments, which is to be further elaborated. It is logical to suppose that if males were offered a female of the same strain, the behavioral and hormonal reaction would be found. Without excluding a standard tester, to avoid multiple interpretations in such experiments it is recommended to investigate also the reaction of males to a receptive female of the same strain, whose most psychophysiological parameters will be the same as those of the male. If males react to a receptive female of the same strain, this will explain the absence of reaction to a receptive female of another strain.

Dynamic Changes of Testosterone Level

Different hormonal response in males to the presence of a receptive female could be due to different dynamics of blood testosterone changes conditioned by a great number of physiological mechanisms. In males of some strains testosterone level may rise twenty minutes (or earlier) after the exhibition of a receptive female [Amstislavskaya, 2006; Amstislavskaya and Khrapova, 2002] and in half an hour not differ from the level in control males. In males of other strains this process may begin later and testosterone enhancement is observed after 40 minutes of exposure [Naumenko et al., 1983]. This is why detailed investigations are needed to obtain a dynamic picture of testosterone level changes to make an adequate conclusion on high or low reactivity of the males of a particular strain to the odor and presence of a receptive female.

Effects of Pharmacological Treatment and Stress

Obviously, any pharmacological preparation under system administration can influence many forms of animal behavior and physiological status. This refers in particular to those preparations that change psychoemotial state, say, by activating or inhibiting neurotransmitter systems of the brain. In such cases it is very difficult to identify specific pharmacological effect on this or that form of behavior. In the studies of sexual motivation it was shown that stimulation and blocking of serotonin receptors of different types [Amstislavskaya and Popova, 2004; Popova and, Amstislavskaya, 2002a; Popova and,

Amstislavskaya, 2002b] most frequently led to a reduction of both the hormonal and behavioral components of the response to a receptive female (although to a different degree). One-way change of the parameters studied could be evoked by changes in the activity of serotonergic system, which is commonly associated with, for example, the development of anxiety [Rodgers and Cole, 1994], which is capable as such of suppressing the behavioral response and testosterone level of the male at the exhibition of a receptive female. That is why such experiments fail to provide unequivocal explanation of what actually happened: whether anxiety (fear) has grown or the sexual motivation has lowered. An increase of anxiety could be a reason of the changes of behavioral and hormonal response of the male under chronic stress, which is probably observable at daily restriction in prenatal age [Kuznetsova et al., 2006]. Additional experiments are needed to understand the specific character of changes in sexual motivation or this phenomenon has to be explained with the use of literature data.

The Features of Interpretation of Behavioral Parameters in the Partition Test

Various approaches to the study, possible experimental designs with the use of the partition test are described in the review article published by the author [Kudryavtseva, 2003]. However, new data have been accumulating on the use of the test in other laboratories, which seems not always to be interpreted correctly. The partition test is actually simple, but the main difficulty is not in measuring the number of times a male "sticks his nose and forepaws" into the partition [Popova et al., 1998] but to interpret behavioral parameters. It is tempting to suggest that the number of approaches reflects motor activity and the duration of stay at the partition – the level of some motivation. However, the number of approaches, obviously, might reflect motor activity only when the neighboring compartment is vacant and clean and the male under investigation has explored the territory of his compartment for not less than a day and passed the period of activation before the test. After the appearance of a partner behind the partition there arises a motivational component of behavior so the assertion that the number of approaches reflects motor activity is to be verified in the tests expressly measuring motor and exploratory activity. Sometimes a clue to understanding is in the derivative parameter of average time spent near the partition on one approach.

Let us imagine the following situation. It was shown that under exposure (for instance, administered preparations) the number of approaches to the partition decreases with the total time of reaction to a partner in the neighboring compartment not changing significantly. If the study deals with the response of male to receptive female the result could be interpreted by a reduced motor activity without influence on the motivation. However if you calculate the average time of one approach by dividing the total time spent near the partition into the number of approaches, it may increase as compared with control males. This means that a male approaches the partition and does not retreat from it craving for the female, while moving continuously at the partition without being distracted for walks around the cage. This points not to a decrease in motor activity but to the enhancement of sexual motivation. One more example is when the number of approaches increases while the average time per one approach significantly decreases as compared with control males. In this case you would rather suggest not an increase in motor activity but the development of two opposite motivations. A mouse first comes to the partition then retreats from it. This "approach-avoidance behavior" is driven by the urge to explore a new object and by fear before the unknown [Montgomery, 1955]. In psychiatry this phenomenon, if it occurs in humans in an acute form, is called ambivalence, meaning that the same incentive is attractive and aversive at the same time. A significant decrease of the average time spent near the partition per one approach was observed in our study of depressive mice [33]. Consideration of only two parameters of the "partition" test narrows the scope of possible interpretations and, thus, the understanding of the observed phenomena.

CONCLUSION: EXPERIMENTAL APPROACHES TO THE STANDARDIZATION OF EXPERIMENTS IN THE STUDY OF MECHANISMS OF SOCIAL INTERACTIONS

The point here seems to be about the obvious thing, standardization of experiments. It is generally accepted that in the experiments the situation for animals is to be maximally equalized, i.e. conditions for control and experimental animals should be the same, excluding the incentive the response to which is being studied. For example, in the experiments measuring the response of animals to a pharmacological treatment control and experimental animals are administered a vehicle and drug, respectively. If animals with different physiological status are investigated, the rule is to equalize the incentive the response to which is being studied. Let it be called a conventional approach.

However common rules of experiment standardization go to pieces when one speaks about social behavior, which is understood as any interaction of two or more individuals on any occasion. Twenty years ago a fact was first confronted (verified later many times) that one and the same stimulus influences differently the behavior and physiological state of animals with different social status. If a finger is brought to the nose of a male with the experience of aggression he may bite, a submissive individual with the experience of social defeats (victim) may run away and an intact animal - exhibit exploratory behavior by sniffing it. This means that motivational and physiological mechanisms of the above behavioral patterns are different and that in individuals with different social status the same stimulus results in entirely different consequences. Similar results were obtained in the study of the influence of different forms of psychoemotional stress on males and females in the work by D. F. Avgustinovich and I. L Kovalenko: social stimuli that induced severe stress in males causing the development of psychoemotional disorders produced no effect on females [Kovalenko, 2006]. The same happens in human life – what is important for one individual is insignificant for another. As for the subject matter, one and the same incentive, receptive female, may activate totally different motivational components of behavior in naïve and experienced males by inducing totally different physiological and behavioral mechanisms.

Therefore, in some experiments it is necessary to search specific incentives (sometimes they are different) producing similar effect on individuals with different emotional status or on male and female, thereby equalizing not the incentive as such but the condition it induces. This means that incentives might be different but their effect - similar. Obviously, to induce sexual motivation in males of different strains you need an appropriate incentive – odor or presence of female. For humans, an appropriate incentive is, in the first place, a visual image of sexual partner. In both cases sexual motivation arises following the same physiological rules. Let us call the approach of using an adequate incentive "adequate", which, as shown by experience, is preferable for the investigation of social behavior. Moreover, it is to be noted that taking an adequate approach may require not so much the use of different incentives for inducing a similar response as totally different conditions to find these or those physiological and psychoemotional peculiarities of each particular individual. For example, in the studies of sexual arousal mechanisms in males with repeated experience of aggression they are to be placed in the conditions that release them from aggressive motivation [Amikishieva and Ovsiukova, 2004], after which they are capable of reacting to a receptive female adequately. In animals with repeated experience of social defeats receptive female might evoke a behavior at the partition similar to control [Kudryavtseva et al., 2004] but it is insufficient to conclude that sexual behavior of such males is intact since social stress is known to lead to the development of gonadal hypofunction [Kudryavtseva et al., 1994] and reproduction impairment [Kaledin et al., 1993]. Additional experiments are needed to say whether the stress of social confrontations affects sexual arousal and, if so, how the affected sexual arousal influences sexual behavior as such. For this purpose detailed investigation of sexual behavior is needed.

Summing up, it may be concluded that in the studies of social behavior for adequate interpretation of the behavioral phenomena are a properly elaborated experimental context and the understanding of physiological regulation of the phenomenon under study, which allows unequivocal interpretation of the motivational component of a behavioral pattern.

REFERENCES

A. Unconditioned sexual incentive motivation in the male Norway rat (Rattus norvegicus). *J Comp Psychol.* 2003. V.117, N 1. P. 3-14.

Alexander G.M., Packard M.G., Hines M. Testosterone has rewarding affective properties in male rats: implications for the biological basis of sexual motivation.// *Behav. Neurosci.* 1994. V.108, N 2. P. 424-428.

Amikishieva A.V. Ovsiukova M.V. The effect of chronic experience of victories or defeats in social conflicts upon sexual motivation in male mice. *Ross Fiziol Zh Im I M Sechenova.* 2004. V.90, N 6. P. 811-819. Russian.

Amstislavskaya T.G. Analysis of behavioral and neuroendocrinic mechanisms of sexual activation of male: role of disadvantage influence in the different periods of ontogenesis and brain serotonin. Dr.Sci. Thesis. 2006, Institute of Physiology, *SD RAMS,* 37 p. Russian.

Amstislavskaya T.G., Khrapova M.V. Effect of genotype on behavioral and hormonal components of sexual activation of male mice. *Bull Exp Biol Med.* 2002. V.133, N 5. P.548-551. Russian.

Amstislavskaya T.G., Osadchuk A.V., Naumenko E.V. Pathways of activation and change of the endocrine function of testes elicited by effect of presence of the female. *Neurosci Behav Physiol.* 1990. V 2. N 6, P. 549-552.

Amstislavskaya T.G., Popova N.K. Female-induced sexual arousal in male mice and rats: behavioral and testosterone response. *Horm. Behav.* 2004.V.46. P. 544- 550.

Amstislavskaya T.G., Popova N.K. The roles of different types of serotonin receptors in activation of the hypophyseal-testicular complex induced in mice by the presence of a female. *Neurosci Behav Physiol.* 2004 V.34. N8; P. 833-837.

Bahrke M.S., Yesalis C.E. 3rd, Wright J.E. Psychological and behavioural effects of endogenous testosterone and anabolic-androgenic steroids. An update. *Sports Med.* 1996. V.22, N 6. P. 367-390.

Bakshtanovskaya I. V., Khodakova E. M., Characteristics of the Responses of the Male Reproductive System on the Actions of Chronic Emotional Stress. *Scientific Studies at the Institute of Problems of Acclimation to the North, Tyumen',* (1992), pp. 25–35.

Berridge K.C. Food reward: Brain substrates of wanting and liking. *Neurosci. Biobehav. Rev.* 1996. N 20. P. 1-25.

Berridge K.C. Motivation concepts in behavioral neuroscience. *Physiol. Behav.* 2004. N 81. P. 179-209.

Bondar' N.P., Kudryavtseva N.N. Impaired social recognition in male mice with repeated experience of aggression. *Zh Vyssh Nerv Deiat Im I P Pavlova.* 2005. V.55, N3. P. 378-384. Russian.

Bronson F. H., Eleftheriou B. E.. Adrenal response to fighting in mice: Separation of physical and psychological causes. *Science.-* 1965. N 147. P. 627-628.

Bronson F.H., Eleftheriou B.E., Dezell H.E. Strange male pregnancy block in deermice: prolactin and adrenocortical hormones. *Biol. Reprod.* 1969. V.1, N 3. P. 302-306.

Bronson F.H., Stetson M.H., Stiff M.E. Serum FSH and LH in male mice following aggressive and nonaggressive interaction physiology and behavior. *Physiol Behav.* 1973. Vol.10, N 2. P. 369-372.

Bruce H.M. A block to pregnancy in the mouse caused by the proximity of strange males. *J. Reprod. Fertility.* 1960. N1. P.96-103.

Bulygina V.V., Amstislavskaia T.G., Maslova L.N., Popova N.K. Effect of chronic stress in prepubertal period on sexual arousal in the male rats. *Ross Fiziol Zh Im I M Sechenova.* 2001. V.87, N7. P. 945-952. Russian.

Clark A.S., Lindenfeld R.C., Gibbons C.H. Anabolic-androgenic steroids and brain reward. *Pharmacol. Biochem. Behav.* 1996. V. 53, N 3. P. 741-745.

Everitt B.J. Sexual motivation: a neural and behavioural analysis of the mechanisms underlying appetitive and copulatory responses of male rats. *Neurosci. Biobehav. Rev.* 1990. V. 14, N 2. P. 217-232.

Finlayson G., King N., Blundell J.E. Is it possible to dissociate 'liking' and 'wanting' for foods in humans? A novel experimental procedure. *Physiol Behav.* 2007 Jan 30;90(1):36-42.

Heiblum R., Arnon E., Gvaryahu G., Robinzon B., Snapir N. Short-term stress increases testosterone secretion from testes in male domestic fowl. *Gen. Comp. Endocrinol.* 2000. V.120, N 1. P. 55-66.

Kagan J. Differential reward value of incomplete and complete sexual behavior. *J. Comp. Physiol. Psychol.* 1955. V.48, N 1, P. 59-64.

Kaledin V.I., Kudryavtseva N.N., Bakshtanovskaia I.V. Anxiety as a possible cause for sex ratio disturbance in a generation ("the war years phenomenon") *Dokl Akad Nauk.* 1993 . V.329, N1. P. 100-102. Russian.

Kovalenko I.L. Gender features of reaction to psychoemotional influence and antidepressants correction in mice of C57BL/6J. *Ph.D. Thesis.* 2006. Novosibirsk, Institute of Cytology and Genetics SD RAS, 16 p. Russian.

N.N. Peculiarities in forming agonistic behavior in mice using a sensory contact model. Review *Novosibirsk*: Institute of Cytology and Genetics SD RAS. 1987. 39 pp. Russian.

Kudryavtseva N.N. Differences in the reactivity of 2 mouse genotypes to zoo-social signals in the partition test. *Zh Vyssh Nerv Deiat Im I P Pavlova.* 1987. V.37, N 5. P. 929-934. Russian.

Kudryavtseva N.N. Behavioral correlates of aggressive motivation in male mice. *Zh Vyssh Nerv Deiat Im I P Pavlova.* 1989. V.39, N 5. P. 884-889. Russian.

Kudryavtseva N.N. Experience of defeat decreases the behavioural reactivity to conspecifics in the partition test. *Behav. Proc.* 1994. N 32. P. 297-304.

Kudryavtseva N.N. The sensory contact model for the study of aggressive and submissive behaviors in male mice. *Aggress. Behav.* 1991. V. 17, N 5. P. 285-291.

Kudryavtseva N.N. Use of the "partition" test in behavioral and pharmacological experiments. *Neurosci Behav Physiol.* 2003, V.33. N 5. P. 461-471.

Kudryavtseva N.N. Amstislavskaya T.G., Kucherjavuy S. Effects of repeated aggressive encounters on approach to a female and plasma testosterone in male mice. *Horm. Behav.* 2004. V. 45, N 2. P. 103-107.

Kudryavtseva N.N., Avgustinovich D.F., Kovalenko I.L., Bondar' N.P. Development of anhedonia under negative experience of social confrontations in male mice. *Ross Fiziol Zh Im I M Sechenova.* 2006. V.92, N 3. P.351-361. Russian.

Kudryavtseva N.N., Bakshtanovskaia I.V., Avgustinovich D.F. The effect of the repeated experience of aggression in daily confrontations on the individual and social behavior of male mice. *Zh Vyssh Nerv Deiat Im I P Pavlova.* 1997. V. 47, N 1. P.86-97. Russian

Kudryavtseva N.N., Bondar N.P., Alekseyenko O.V. Behavioral Correlates on Learned Aggression. *Aggress. Behav.* 2000. V. 26, N 5. P. 386-400.

Kudryavtseva N.N., Bondar N.P., Avgustinovich D.F. Association between repeated experience of aggression and anxiety in male mice. *Behav. Brain Res.* 2002. V.133, N 1. P. 83-93.

Kudryavtseva N.N., Koriakina L.A., Sakharov D.G., Serova L.I. The effect of the prolonged experience of aggression and subordination on the adrenal cortical and androgenic functions of inbred male mice. *Fiziol Zh Im I M Sechenova.* 1994. V.80, N 11. P. 26-31. Russian.

Kuznetsova E.G., Amstislavskaia T.G., Bulygina V.V., Popova N.K. The stress effect in the prenatal period on sexual excitation and sexual orientation of the mice males. *Ross Fiziol Zh Im I M Sechenova.* 2006. V.92, N1, P.123-132. Russian.

Lopez H.H., Olster D.H., Ettenberg A. Sexual motivation in the male rat: the role of primary incentives and copulatory experience. *Horm. Behav.* 1999. N 36. P.176–187.

Mazur A., Booth A. Testosterone and dominance in men. *Behav. Brain Sci.* 1998. V. 21, N 3. P. 353-363.

Mela D.J. Eating for pleasure or just wanting to eat? Reconsidering sensory hedonic responses as a driver of obesity. *Appetite.* 2006. V. 47, N 1. P. 10-17.

Menning O. *Animal Behavior.* M. MIR,1982, 360p

Montgomery, K.C. The relation between fear induced by novel stimulation and exploratory behavior. *J. Comp. Physiol. Psychol.* 1955. N 48. P. 254-260.

Naumenko E.V., Amstislavskaya T.G., Osadchuk A.V. The role of adrenoceptors in the activation of the hypothalamic-pituitary-testicular complex of mice induced by the presence of female. *Exper. Clinic.Endocrin.* 1991. V.97, N 1. P.1-12.

Naumenko E.V., Osadchuk A.V., Serova L.I. Genetic and neuroendocrine mechanisms underlying the control of hypothalamic-hypophysial-testicular complex during sexual arousal. *Soc. Sci. Rev. F. Physiol. Gen. Biol. 1992.* N 5. P. 1-39.

Naumenko E.V., Osadchuk A.V., Serova L.I., Shishkina G.T. Genetic-physiological mechanisms of testicular function regulation. Novosibirsk, *Nauka,* 1983, 201 p.

Novikov S. N., Pheromones and Mammalian Reproduction, *Nauka,* Leningrad 1988. 132 p.

Osadchuk A.V., Naumenko E.V. Role of the genotype and zoosocial behavior types in regulating testicular endocrine function in mice. *Dokl Akad Nauk SSSR* 1981. V. 58, N 3. P. 746-749. Russian.

Pfaus J.G., Kippin T.E., Centeno S. Conditioning and sexual behavior: a review. *Horm. Behav.* 2001. V.40, N 2. C. 291-321.

Popova N.K., Amstislavskaia T.G., Kucheriavyi S.A. Sexual motivation in male mice induced by the presence of the female. *Zh Vyssh Nerv Deiat Im I P Pavlova.* 1998. V. 48, N 1. P. 84-90. Russian.

Popova N.K., Amstislavskaya T.G. 5-HT2A and 5-HT2C receptors differently modulate mouse sexual arousal and the hypothalamo-pituitary-testicular response to the presence of a female // *Neuroendocrinology.* 2002a.V.76, N 1. P. 28-34.

Popova N.K., Amstislavskaya T.G. Involvement of the 5-HT(1A) and 5-HT(1B) serotonergic receptor subtypes in sexual arousal in male mice. *Psychoneuroendocrinology.* 2002b. V. 27, N 5. - P.609-618.

Portillo W., Paredes R.G. Sexual incentive motivation, olfactory preference, and activation of the vomeronasal projection pathway by sexually relevant cues in non-copulating and naive male rats. *Horm. Behav.* 2004. V. 46, N 3. P. 330-340.

Robinson T.E., Berridge K.C. The neural basis of drug craving: an incentive-sensitization theory of addiction. *Brain Res Brain Res Rev.* 1993. V.18, 3. P. 247-291

Robinson T.E., Berridge K.C. The psychology and neurobiology of addiction: an incentive-sensitization view. *Addiction.* 2000. V.95, Suppl 2. P. S91-117

Rodgers R.J., Cole J.C. The elevated plus-maze: pharmacology, methodology and ethology. In Cooper SJ and Hendrie CA (eds.), *Ethology and Psychopharmacology,* John Wiley & Sons Ltd., Chichester, 1994. pp. 9-44.

Rubinow D.R., Schmidt P.J. Androgens, brain, and behavior. *Am. J. Psychiatry.* 1996. V. 153, N 8. P. 974-984.

Simonov P.V. Motivated brain. M., 1987.

Steel E. Male-female interaction in the hamster: seasonal and hormonal effects. In: *Hormones and Behaviour in Higher Vertebrates.* J.Balthazart, E.Prove, R.Gilles (eds). Springer-Verlag Berlin Heidelberg: 1983. P. 314-324.

Suay F., Salvador A., Gonzalez-Bono E., Sanchis C., Martinez M., Martinez-Sanchis S., Simon V.M., Montoro J.B. Effects of competition and its outcome on serum testosterone, cortisol and prolactin. *Psychoneuroendocrinology.* 1999. V.24, N 5. P. 551-566

Whalen R.E. Effects of mounting without intromission and intromission without ejaculation on sexual behavior and maze learning. *J. Comp. Physiol. Psychol.* 1961. N 54. P. 409–415.

Wood R.I. Reinforcing aspects of androgens. *Physiol Behav.* 2004. V.83, N 2. P. 279-289.

In: Animal Behavior: New Research
Editors: E. A. Weber, L. H. Krause

ISBN: 978-1-60456-782-3
© 2008 Nova Science Publishers, Inc.

Chapter 4

GAZE FOLLOWING IN NON-HUMAN ANIMALS: THE CORVID EXAMPLE

Christian Schloegl[1,2], Judith Schmidt[1,2], Christelle Scheid[3], Kurt Kotrschal[1,2] and Thomas Bugnyar[1,4]

[1]Konrad Lorenz Forschungsstelle, Grünau im Almtal, Austria
[2]Department of Behavioural Biology, University of Vienna, Austria
[3]IPHC-DEPE, ULP, CNRS, Strasbourg, France
[4]Centre for Social Learning and Cognitive Evolution, School of Psychology, University of St. Andrews, UK

ABSTRACT

By definition, following others' gaze occurs, if an individual shifts its attention towards the visual focus of others. Originally considered to be an unique human skill, this view has been challenged by pioneering work of Povinelli & Eddy (1996), who showed gaze following abilities in chimpanzees (*Pan troglodytes*). Since then, many mammalian species have been found to use others' gaze as a source of information with apes being capable to follow gaze geometrically behind visual barriers. This suggests that they may even understand how barriers impair their own visual perception. Taken together, the data suggest that the use of gaze cues may represent an important skill to obtain information in social mammals. However, hardly anything is known about such skills in non-mammalian species, rendering the evolutionary history of gaze following unclear.

With respect to birds, our knowledge is limited to a single species, the common raven (*Corvus corax*). In order to gain an evolutionary perspective, we here present data on two closely related corvids, rooks (*C. frugilegus*) and jackdaws (*C. monedula*). As ravens, the food-caching rooks appear to be capable of following others' gaze behind barriers; additionally, the temporal pattern of the ontogenetic development of gaze following is similar in both species. In contrast, we found no evidence that the non-caching jackdaws follow human gaze, but have some indication that they may follow conspecifics' gaze. An understanding of barriers, however, could not be demonstrated in jackdaws. These findings support the idea that geometrical gaze following and gaze following into distant space might serve different functions. We suggest that gaze following into distant space may primarily reflect an anti-predator response, whereas geometrical gaze following skills may have developed to protect food caches from scroungers. Finally, we will discuss alternative explanations and will present an outline for potential future research.

INTRODUCTION

Responsiveness to gaze is an interesting phenomenon for comparative cognitive research, because gaze following may require an understanding of what other can see, but oneself does not (for reviews see Emery 2000; Itakura 2004). Baron-Cohen (1995) considered gaze following to be one building-block for the development of a human-like "Theory of Mind" (Premack and Woodruff 1978). Due to its potential importance for understanding human cognitive abilities, gaze following was predominantly studied by developmental psychologists (e.g. Baron-Cohen 1995; Brooks and Meltzoff 2005; Butler et al. 2000; Butterworth and Jarrett 1991; Flavell et al. 1981). Butterworth and Jarrett (1991) investigated the development of human gaze following in great detail: when watching a human ('model') looking at an object, children at six months of age change their viewing angle to look at this object only if the object was in their visual field before. Between 12 to 18 months of age, children start to turn around if the model gazes at a target behind the child's back. At 18 months of age, children move, walk or crawl around a barrier, if it obstructs their view from what the model is looking at (but see also Moll and Tomasello 2004). Studying gaze following in non-human animals is a rather recent development in cognitive science.

Due to ambiguities in terminology, we will employ certain technical terms as follows: "Gaze" is most frequently used to describe head and eye – orientation, whereas "glance" is used to describe a cue based on eye-direction only. Although some animals seem to be responsive to the eye-direction of others (e.g. jackdaws, ravens and rooks; von Bayern, pers. communication, Bugnyar & Schloegl unpubl. data), most studies on non-human animals focused on the responsiveness to head and eye – direction (e.g. Bräuer et al. 2005; Kaminski et al. 2005). In birds, a distinction between gaze and glace may not prove useful, since birds usually turn their head and eyes simultaneously (Dawkins 2002). Consequently, the term "gaze" is used to describe the combined orientation of the head and the eyes. Secondly, we use "gaze following" in the sense of "looking at what others are looking at" (e.g. Corkum and Moore 1995; Tomasello et al. 1998), which others had termed "visual co-orientation" (e.g. Anderson and Mitchell 1999) or "joint visual attention" (e.g. Butterworth and Jarrett 1991).

THE COGNITIVE BASIS OF GAZE FOLLOWING

It has been suggested that looking into the same direction as others, often referred to as "gaze following into distant space", (e.g. Schloegl et al. 2007) and following other's gaze around barriers ("geometrical gaze following"; Tomasello et al. 1999) may reflect two different "modes" of gaze following (Gomez 2005). These different modes may reflect different cognitive mechanisms (Bugnyar et al. 2004). Povinelli & Eddy (1996) proposed a "low-level" and a "high-level" model, the former representing a socially facilitated orientation response, which causes an individual to look in the same direction as others. This "low-level" mechanism, however, may not be sufficient to explain geometrical gaze following, because then individuals should look at the barrier itself instead of moving around it. Therefore, Povinelli & Eddy proposed a "high-level" mechanism, which requires an individual to understand that model and observer do not see the same things, because the barrier affects their visual perception differently. This does not necessarily require a mental representation of the other's perception, but may be achieved by associative learning (Tomasello et al. 's (1999) 3rd – level explanation). Accordingly, a "low-level" mechanism may be sufficient to follow gaze geometrically, if the individual possesses the capacity to learn how barriers may or may not impair an individual's perception.

GAZE FOLLOWING IN MAMMALS

In the mid-1990s', comparative psychologists became interested in gaze following skills in non-human animals. Povinelli and Eddy (1996) were the first to report gaze following in chimpanzees (*Pan troglodytes*). When observing a human model looking to the left, to the right or up, chimpanzees changed their orientation and looked in the same direction. The animals also followed the human model's gaze behind a distractor, as though they understand whether an object blocks their view. Further on, Tomasello et al. (1999) indeed demonstrated geometrical gaze following skills in chimpanzees, which has since been expanded to all great apes (Bräuer et al. 2005). Gaze following into distant space also has been reported in a variety of monkeys (Anderson and Mitchell 1999; Emery et al. 1997; Tomasello et al. 1998). Positive results in goats (*Capra hircus*; Kaminski et al. 2005) and potentially, dogs (*Canis familiaris;* Bräuer et al. 2004) suggest, that gaze following may not be a particular skill of primates, but may be found in other mammalian taxa.

"FORGOTTEN" QUESTIONS ABOUT GAZE FOLLOWING: FUNCTIONAL VALUE AND EVOLUTIONARY HISTORY

Most studies discuss gaze following as a tool to use others as a source of information about predators, food and/or important social interactions (e.g. Tomasello et al. 1999). However, geometrical gaze following has been reported only for large-brained apes and ravens, but not for monkeys and goats. It remains unclear, whether this is due to a lack of testing, to the use of inappropriate methodology, or whether it truly reflects species-specific differences as of yet not revealed due to a publication bias. Given that differences between species exist, we would expect differential selection pressure on the two modes of gaze following, which in turn would favour the idea of different functional backgrounds. We investigated this idea in our most recent research in ravens. First, Bugnyar et al. (2004) demonstrated that ravens pass the standard tests for gaze following into distant space as well as geometrical gaze following behind barriers. After a raven saw a human foster parent look up, it also turned its head to look up. Similar, after the bird saw the foster parent look behind a barrier, it walked around the barrier or flew on top of it, presumably to discover what was behind the barrier (see Figure 1 for a photo of the same experimental setup used in rooks). In a second study, we investigated the ontogenetic development of gaze following in ravens. We tested hand-raised ravens once a week to determine at what age they followed a human model's look-up, starting two weeks pre-fledging (Schloegl et al. 2007). As soon as the birds were capable of following human look-ups, we tested once a month, whether the birds would move around a barrier after they saw the human model looking behind it (Figure 1). The birds started to follow human look-ups within the first weeks post-fledging regularly, but did not follow human gaze cues behind barriers at this early age. Geometrical gaze following skills were first detectable when the birds were between seven and eight months old.

Interestingly, the emergence of both skills coincided with important steps in the development of young ravens. At the time around fledging, when ravens start to follow human look-ups, they become vulnerable to avian predators or approaching competitors. Gaze following into distant space may now be a useful tool. In fall, juvenile birds become fully independent from their parents (Haffer and Kirchner 1993; Heinrich 1989), integrate in the non-breeder flock (Heinrich 1989) and have to find food for themselves. At that time, juvenile ravens' caching behaviour changes: birds begin to cache out of sight of conspecifics more regularly than before (Bugnyar et al. 2007). For caching out of sight, ravens may rely on a more complicated cognitive mechanism than "out of sight – out of mind", but may be aware of what potential pilferer can and cannot see (Bugnyar and Heinrich 2005, Bugnyar and Heinrich 2006). On the other hand, we failed to demonstrate that ravens rely on humans' or conspecifics' gaze cues to detect unknown food caches (Schloegl et al. 2008a; Schloegl et al. 2008b) This led to the idea that geometrical gaze following may represent a by-product of ravens' understanding of the function of barriers, which may be required for caching out of sight (Bugnyar et al. 2004; Schloegl et al. 2007).

Figure 1. Experimental setup for testing rooks' ability to follow human gaze behind visual barriers. In the foreground, the experimenter's head is visible while she looks behind the barrier.

Great apes do not cache food, but follow gaze geometrically as well. We propose that the functional value still may be similar: primates, and apes in particular, are well-known for their deceptive behaviour (e.g. Byrne and Whiten 1988), often concealing their actions from conspecifics by going out of sight. An advantage of deception may have led to the development of an understanding for barriers. Although intuitively compelling, this hypothesis is so far based only on a limited number of species studied, and more comparative data from a wide array of species are needed.

CORVIDS AS A MODEL SPECIES FOR COMPARATIVE STUDIES

Birds and mammals are interesting taxa for comparative studies and thus for deciphering evolutionary scenarios. Both groups developed independently from a common reptilian ancestor, which lived at least 300 million years ago. Consequently, similar cognitive skills in both taxa indicate either ancient heritage or convergent evolution of similar abilities. The latter is of particular interest, because it may provide insight into the driving forces of cognitive evolution. Whereas mammals, primates, cetaceans, elephants and canids are of particular interest, two bird clades are the main focus: parrots and corvids. Both groups possess brain sizes comparable to primates (Emery 2006; Emery and Clayton 2004) and developed a "cognitive cerebro-type" (Iwaniuk and Hurd 2005). This indicates that brain areas, which are particularly important for advanced cognitive skills, are enlarged.

Similar to apes, corvids primarily are orientated visually and appear to be highly attentive to others' perceptions, i.e. ravens regularly pilfer food caches of conspecifics and rely on the observation of the caching event for successful pilferage (Heinrich and Pepper 1998). Additionally, they are long-living, have prolonged juvenile periods and most species live in complex social units. For example, subadult ravens join fission-fusion groups, in which they find their life-long pair partners, before they reach sexual maturity (e.g. Heinrich 1989; Heinrich 1999). Rooks and jackdaws spend their entire life in large groups and breed in colonies. These similarities make corvid species suitable for comparative research.

STUDYING GAZE FOLLOWING IN CORVIDS

We conducted experiments similar to those done in ravens, with several other corvid species. Initially, we used two species closely related to ravens, but with some differences in their socio-ecological organization: similar to ravens, rooks regularly cache food, although their caching behaviour appears to be more seasonal; additional, rooks are socially more tolerant than ravens (Madge and Burn 1994). Rooks are facultative migratory, whereas ravens are not. Ravens, however, wander regularly over large distances between different non-breeder flocks (e.g. Heinrich et al. 1994). Jackdaws have a social organization similar to rooks, but are one of only two non - food caching corvid species (de Kort and Clayton 2006).

TESTING GAZE FOLLOWING IN ROOKS

In 2006, we (J. Schmidt, C. Scheid) hand-raised a group of 16 rooks. We followed the methodological protocol previously applied to ravens to study the development of the rooks' gaze following skills (Schloegl et al. 2007). Birds were taken from nests when approx. 12 – 30 days old. During the daily handling routine of the birds, we collected the following data:

first, the hand-raisers noted *ad libitum* whenever one bird turned its head approx. 90° to look up one eyed. This is the typical behaviour of birds looking up (Dawkins 2002). Additionally, they recorded if a second bird looked up within five seconds after the look up of the first bird. Although only suggestive, these occurrences of co-orientation may represent gaze following events.

One week before the estimated time of fledging (at about 30 - 35 days of age), birds were experimentally tested for their ability to follow human look-ups. Pre-fledging, birds were tested when they were separated from their nest-mates during daily cleanings of the nest boxes. We used these regular cleaning intervals, because birds were habituated to this temporal separation from their nest-mates. Post-fledging, birds were isolated temporarily from conspecifics somewhere inside the aviary, where the experimenter positioned herself approx. 1m away from the bird. Birds received their first session three days post-fledging, and from there on were re-tested every second week. During each session, birds received only one test and one control trial. At the onset of a trial, the experimenter called the bird's name to gain its attention. As soon as the bird faced the experimenter, she either looked up (test trials) or to a previously designated point behind the bird (control trial) for five seconds. Later on, we coded from videotapes whether the birds looked up within fifteen seconds after the onset of the cue.

In a consecutive experiment, we tested the birds for their ability to follow human gaze cues behind visual barriers once per month (see also Schloegl et al. 2007). To increase temporal resolution, we divided birds in two groups. One group was tested first in their 12th or 13th week post-fledging, whereas the second group was tested first in the 14th or 15th week post-fledging. From then on both groups were tested every fourth week. After a statistically detectable difference between test and control trials was found, all birds received one additional session within two weeks.

Birds were tested individually, with a brick-wall (approx. 1 m high and 1.5 m in length) functioning as visual barrier (Figure 1). As soon as the bird positioned itself appropriately on one side of the barrier, the experimenter looked either to the other side of the barrier (test trial) or to a point behind the bird (control trial). Otherwise, trials were identical to look-up trials. Similar to the raven study, participation in the experiments was voluntarily and, therefore, the number of birds varied between sessions, depending on their motivation.

In contrast to our raven studies, we introduced one important methodological difference: we always used the same experimenter as model to test the ravens, but our results suggest that the birds habituated to the setup very quickly (Schloegl et al. 2007) and an exchange of the experimenter was necessary to elicit a response. Therefore, from the start of the experiment, we used three different models for tests with rooks. All models were equally familiar to the birds and each experimenter was used as a model in a different week.

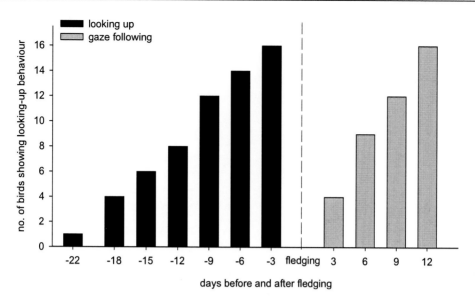

Figure 2. Cumulative number of rooks showing one-eyed looking up (black bars) and gaze following (grey bars) during daily observations. Bars represent 3-day blocks, the vertical dashed line represents day of fledging.

The results obtained from rooks closely resembled the results from ravens. In the pre-fledging period we did not record a single case of a bird looking up following the look up of another bird (Figure 2). However, this changed after fledging: both hand-raisers began to observe some of these events and within twelve days post-fledging all birds were seen at least once to follow another rooks' gaze. This is in concordance with the results obtained from ravens and supports the experimental data: when the birds were tested during the last week before fledging as well as a three days post-fledging, there was no different response between test and control trials (McNemar tests: pre-fledging: N = 8, P > 0.999; post-fledging: N = 10, P > 0.999; Figure 3). Starting in the second week post-fledging, birds looked up more frequently in test-trials relative to control trials (McNemar: N=11, P=0.031; Fig 3). As soon as the rooks exhibited gaze following abilities, they seemed to habituate and stopped responding in subsequent test sessions (McNemar tests: all P > 0.25; Figure 3).

We obtained a similar result when birds were tested for geometrical gaze following skills. In the initial test sessions, rooks did not respond at all: neither in test- nor control trials did they walk around the barrier. As birds showed no response, a statistical analysis of these sessions is not possible (Figure 4). In the 16th – 17th week post-fledging, however, 67 % of the birds walked around the barrier, if they observed the experimenter looking there (McNemar: N = 9, P = 0.063). Once more, during the following session, birds stopped responding (McNemar: N = 8; P = 0.250), suggesting that they habituated to the setup.

To sum up, gaze following seems not to be limited to ravens, but can be found also in rooks, although the rook results are statistically less powerful than the raven results. The developmental pattern appears to be similar in both species. Gaze following into distant space develops shortly after fledging, with rooks showing a significant response earlier, relative to time of fledging, than ravens. Still, it is unclear, whether rooks develop gaze following skills earlier than ravens. Ravens appeared to habituate very quickly (Schloegl et al. 2007) and this may have masked gaze following skills at the age at which rooks responded. Geometrical gaze following skills seem to develop later in life, approximately at the time at which birds become independent from their parents (Haffer and Kirchner 1993; Madge and Burn 1994). Unfortunately, the ontogenetic development of rooks' caching behaviour has not been studied in enough detail to allow a direct comparison with ravens. Therefore, we cannot answer, whether gaze following skills and food caching develop in parallel in both species.

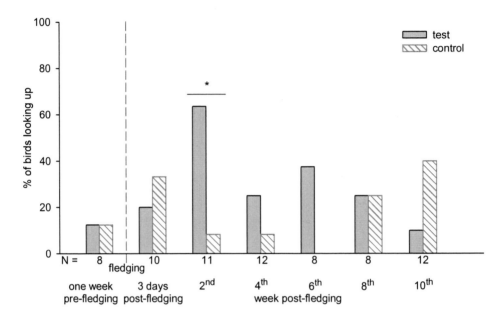

Figure 3. Percentage of rooks looking up when tested for their ability to follow human look-ups. The vertical dashed line represents the day of fledging.

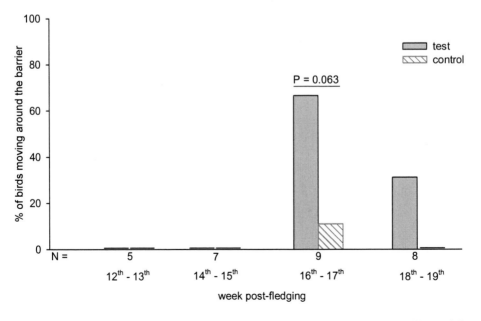

Figure 4. Percentage of rooks moving around a barrier when being tested for their ability to follow human's gaze behind visual barriers.

JACKDAWS

In 2006, we hand-raised a group of jackdaws and performed exactly the same experimental procedure as in rooks, again using three different hand-raisers as models. However, we failed to detect a response in jackdaws (Drew Gaede, unpubl. data). Therefore, we repeated our experiments with another group of thirteen hand-raised jackdaws in 2007, which we took from their nests at 10 - 14 days of age. Initially, we kept these birds in artificial nest boxes with their siblings, but moved the birds to a small outdoor aviary a few days before fledging. Jackdaws leave the nest and hop around in the proximity of the nest prior to being able to fly. Consequently, we defined fledging as the time we first saw a bird fly more than 2 m.

To facilitate testing, we habituated the birds to being removed temporarily from their nest boxes/ aviary. While perched on the experimenter's finger, birds were carried behind a wall to test them in visual isolation. We followed our experimental procedure, using the three hand-raisers as models. Birds received one session every five days, with test and control trials in randomized order. We ran the experiment over a period of twenty days, a time period from shortly before fledging until well thereafter. Again, we did not find any significant difference between test and control trials (all McNemar tests, all N = 13: session 1: P > 0.999, session 2: P = 0.687, session 3: P > 0.999, session 4: P = 0.219). Over the course of the experiment, birds actually appeared to become less interested in the experimenters' behaviour, started to refuse to participate by flying off the finger before testing, and we finally terminated the experiment.

Does our repeated failure to show a gaze following response in jackdaws indicate that they are not capable of following other's gaze, or do these birds simply not follow the gaze of *humans*? Several monkey species have been reported not to follow human gaze cues (e.g. Itakura 1996), whereas some of these species appear to follow conspecifics' gaze (Tomasello et al. 1998). We had speculated previously that ravens may be particularly responsive to heterospecific gaze cues, since they are regularly confronted with heterospecific competitors and predators like wolves, lynx, bears, humans, eagles and buzzards (Schloegl et al. 2007). It is probable that jackdaws may be less prone to follow human gaze than do ravens or rooks. Hence, we re-tested our birds with conspecific models.

TESTS WITH CONSPECIFIC MODELS

Experiments with conspecific models were performed several weeks after the end of the ontogeny study to avoid that the two experiments would influence one another. For individual testing, we trained birds to be temporarily separated from the rest of the group. The test compartment consisted of two adjacent rooms, separated from each other by wire mesh. This allowed that birds in both rooms could observe one another. We attached an opaque curtain approximately half a meter above the birds' head (Figure 5). For each bird, we randomly assigned another bird as model. During tests, one bird was positioned in each room, and as soon as the birds faced each other, the trial began. In total we used three different conditions and each bird received only one condition per day:

Figure 5. Experimental procedure to test the ability to follow conspecifics' look ups.
The black line marks an opaque curtain, the grey line marks a wire mesh partition.
A laser-pointer was used to project a red dot on the opaque curtain above the bird's head.

Test: the human experimenter used a laser – pointer to project a red dot on the curtain above the model's head, until the model turned its head to look up. We considered a response as gaze following if the observer bird looked up within fifteen seconds after the model had first looked up.

Control with another bird present: again, two birds were in place, but the human experimenter did not project the laser pointer's red dot. For the analysis, we randomly picked a 30s-sequence from the video – recording, in which the model bird did not look up. We then checked, whether the observer bird looked up within the last 15 sec of the recording; the first 15 sec were analysed to ensure that the model bird did not provide any gaze cue shortly before the onset of the analysed sequence.

Control alone: this time, no model bird was present but the human experimenter projected for approx. 3s a red dot on the side of the curtain, at which the model bird would have been during the test condition. This control was conducted to ensure that the observer bird could not see the red dot accidentally or that the behaviour of the human experimenter caused the observer bird's response. For the analysis, we measured again, whether the bird looked up within fifteen seconds after the experimenter had projected the red dot.

Birds received the three conditions in randomized order. To avoid habituation, each bird received each session only once and test days were separated by at least one day without further tests. Within each session, each cue was repeated up to three times, with a minimum of 30 sec between each cue. However, this was not always feasible, because models could not be motivated to look up repeatedly or because observer birds left the test compartment. For analysis we used a mean of 2.5 cues per bird for the test condition, 2 cues per bird for the 'control with other bird' – condition and 2.58 cues per bird for the alone – condition.

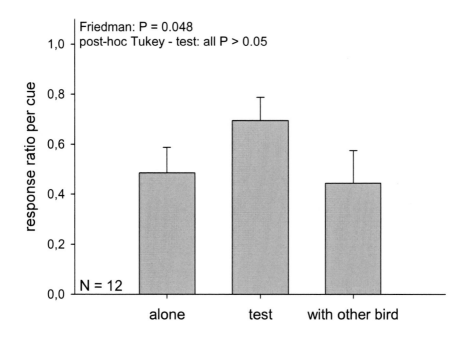

Figure 6. Jackdaws' response to conspecific look-ups. Given is the proportion of cues that led to a response by the birds (i.e. 1 = always responded; 0 = never responded). Bars show x +/- SE. Hence, we failed to demonstrate a clear response by jackdaws to conspecific gaze cues. Nevertheless, we attempted to apply the methodology of using conspecific models, to test if jackdaws would follow gaze cues behind visual barriers. Again, two birds were facing each other while being separated by a wire mesh partition (Figure 7). On both sides of the partition, the room was further divided by a barrier: on the model's side this barrier was made of wire mesh, but on the observer's side the barrier was solid. In other words, the model bird could see behind the barrier on the observer's side, while the observer's view was blocked (see Figure 7). Still, the observer bird could see the model looking behind the barrier. To move around the barrier, observers could either walk around or could fly on top of it to look down. The testing procedure was otherwise the same as previously described: we randomly assigned models to observers and observers received one session per day, with one condition per session and the order of conditions being randomized.

As the number of cues per bird per condition differed, we compared the response to the first cue per condition in a first step, but we did not find any difference between test and control – trials (Cochran's Q: N = 12, Q = 0.667, df = 2, P = 0.717). In a second step, we calculated a proportion of cues, which lead to a response, for all cues given (i.e. # look ups/# of cues) and found a significant difference between the three conditions (Friedman: N = 12, χ^2 = 6.059, df = 2, P = 0.048). The visual inspection of Figure 6 indicated a higher response ratio in test than in control trials, but this did not hold in a post-hoc analysis (Tukey – test; for all pair wise comparisons P > 0.05).

We provided three different conditions:

Test conditions: two birds were present and the experimenter used a laser-pointer to project a red dot on the ground behind the barrier; the dot was visible for model birds only (see Figure 7).

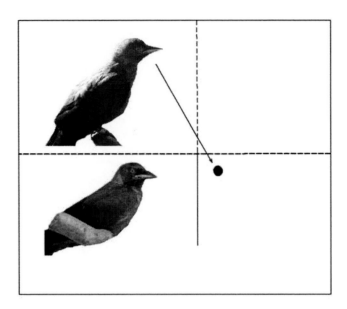

Figure 7. Sketch of the methodology applied to test the response of jackdaws to conspecific gaze cues behind barriers. The dashed line indicates a wire mesh partitions, solid lines indicate opaque partitions. The dot symbolizes the site of the projection of the laser-pointer dot and the arrow indicates the model's gaze direction in the test condition.

Control with other bird present: two birds were present but the experimenter did not project the laser pointer's red dot.

Control alone: only the observer bird was present, but the experimenter projected a red dot behind the barrier.

Again, a response of the observer birds had to occur within 15 sec after the onset of the cue or after the designated starting point of a randomly selected sequence in the 'control with other bird present' condition.

As in the previous experiment, we intended to present each cue three times per session, with an inter-cue interval of at least thirty seconds. Again, not all birds participated long enough to allow the presentation of three cues. Observer birds received a mean of 2.63 cues in the test condition, 2.91 cues in the 'control with other bird' condition and all birds received three cues in the alone condition.

When comparing the response of the first cue in each condition, we could not detect any difference between the three conditions (Cochran's Q: N = 11, Q = 0.4, df = 2, P = 0.819). In contrast to the previous experiment, a comparison of the proportion of cues leading to a response did not provide any significant difference between conditions (Friedman: N = 11, χ^2 = 1.152, df = 2, P = 0.562).

In sum, we could not detect any indication for a response of jackdaws to conspecific's gaze cues behind visual barriers. Also the evidence for gaze following into distant space remains unconvincing and is limited to experiments with conspecific models. If jackdaws are capable of following the gaze of others, our results suggest that their gaze following abilities may be qualitatively and/or quantitatively different from gaze following in rooks and ravens or they cannot be detected with the same experimental tools.

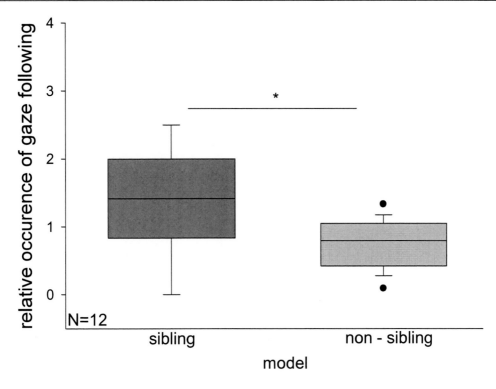

Figure 8. Responsiveness of ravens to gaze cues of siblings and non-siblings during regular observational protocols. To calculate a ratio for the relative occurrence of gaze following, we corrected the number of gaze following responses for the number of siblings and non-siblings (e.g.: no. of gaze following responses to siblings/no. of siblings in the group). Box-plots give median and upper and lower quartile. Whiskers indicate 10% and 90% - confidence interval, dots indicate outliers. To test ravens with conspecific models, we used the methodology previously described for jackdaws (see Figure 5), with some minor modifications. Each bird received two sessions, one session with a sibling as model and the other session with a randomly selected non-sibling as model. Whether birds were first tested with a sibling or with a non-sibling as model was randomized. As it is known that ravens are capable of following others' gaze, we applied the above described test condition only. Since the number of look-ups differed between model birds, we performed two analyses: first, we focused on the first cue provided per model only, and calculated the number of birds, which responded, and secondly, we calculated a response ratio by dividing the number of look-ups per observer bird by the number of look-ups per model bird.

One possible explanation that we could not show gaze following in jackdaws may be that they appear to be more vigilant than rooks or ravens. In test trials, jackdaws did not look up less frequently than ravens or rooks (in all species, birds looked up in approx. 70% of all test trials), but jackdaws looked up more frequently in control trials. This high baseline of vigilance may either mask jackdaws' gaze following abilities, or jackdaws may not follow gaze because it may be less adaptive for highly vigilant birds. It remains speculative, whether this interpretation may also explain the lack of evidence for gaze following in the barrier-task.

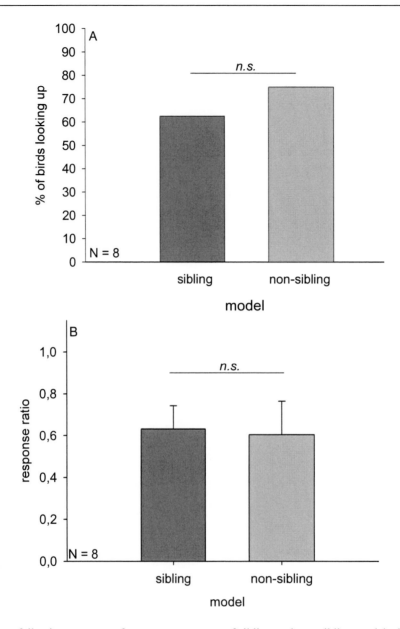

Figure 9. Gaze following response of ravens to gaze cues of sibling and non-sibling models during experimental testing. A. Percentage of birds responding to the first cue of the respective model. B. Response ratio, calculated as number of look-ups by the observer/number of look-ups by the model. Bars show x +/- SE.

GAZE FOLLOWING RESPONSE TO DIFFERENT SOCIAL ALLIES

Most published work on gaze following in non-human animals focused on the question whether animals can or cannot follow the gaze of other, but hardly anything is known beyond that scope. In our long-term monitoring of a group of captive ravens, we conducted daily five minute - focal observations of all birds, collecting a wide range of behavioural data. During the birds' first summer (late April – September 2004; 68.67 +/- 1.79 observations (x +/- SE) per bird), we recorded if two birds looked up ≤ five seconds apart and which birds were involved. Although we cannot distinguish between "true" gaze following events and independent responses of two birds to the same stimulus, these data may still provide insight into the "natural" occurrence of gaze following in ravens. Previous studies showed that ravens pay more attention to closely affiliated individuals than to non-affiliated birds (Scheid et al. 2007), and learn more readily from siblings than from non-siblings (Schwab et al. 2008). Interestingly, the observed cases of visual co-orientation involved sibling pairs more frequently than non-sibling pairs (paired t-Test: N = 12, T = 2.874, df = 11, P = 0.015; Figure 8). In contrast, ravens did not respond stronger to affiliated than to non-affiliated birds (Wilcoxon: N = 12, Z = -1.6, P = 0.110). We also found no influence of rank relationships (paired T-Test: N = 12, T = 1.07, P = 0.308) or sex (paired T – Test: N = 12, T = -1.268, P = 0.231). However, siblings spent more time sitting close together, i.e. within a distance of < 30cm, than non-siblings (paired T-Test: N = 12, T = 4.699, df = 11, P = 0.001). Hence, the higher responsiveness to gaze cues of siblings may be a result of closer spatial proximity. We used an experimental approach to test whether look-ups of siblings would elicit a stronger response than look-ups of non-siblings.

We did not find a difference in gaze following response of ravens to siblings or non-siblings: when focussing on the first look up per model, five out of eight birds responded to siblings and six out of eight birds responded to non-siblings (McNemar: N = 8, P > 0.999; Figure 9A). Similarly, we did not find a difference when comparing response ratios (paired T-test: N = 8, T = 0.116, df = 7, P = 0.911; Figure 9B).

These results suggest that ravens may pay more attention to affiliated than to non-affiliated individuals (Scheid et al. 2007) and may learn more readily from siblings than from non-siblings (Schwab et al. 2008), but this may be limited to specific contexts. Very likely, in an anti-predator context, siblings' and non-siblings' behaviour appears equally salient and ravens are equally likely to follow gaze.

CONCLUSION

What do four years of research on gaze following in corvids tell us? First, we have clear empirical evidence for advanced gaze following skills in corvids. We demonstrated geometrical gaze following skills of three different groups of animals of two different species, rooks and ravens, in three different labs (at the University of Vermont, at the Konrad Lorenz – Forschungsstelle and at the University of Strasbourg). Second, in both species, the ontogenetic development of gaze following is similar. We found evidence for gaze following into distant space shortly after fledging, and geometrical gaze following skills several weeks later. Third, we found only weak evidence for gaze following in jackdaws. Such negative results are difficult to interpret. Either jackdaws do not follow the gaze of others, or it may be difficult to show a gaze following response, since jackdaws appear to express a higher basal vigilance level than rooks and ravens.

The cognitive mechanisms underlying gaze following remain unclear. Povinelli and Eddy (1996) argued that chimpanzees may possess a high-level mechanism, since their low-level model is not sufficient to explain geometrical gaze following. In contrast, Tomasello et al.

(1999) argued against two different models and proposed their 3rd level explanation. Based on the different ontogenetic development of the two modes of gaze following in ravens, we suggested gaze following into distant space and geometrically around barriers to have different underlying cognitive mechanisms (Bugnyar et al. 2004; Schloegl et al. 2007). Our new results of rooks do not rule out that two different mechanisms are involved.

Our results may have some important implications towards an understanding of the selection pressures driving the evolution of gaze following skills. Gaze following into distant space is found in a variety of species, including apes (e.g. Bräuer et al. 2005; Povinelli and Eddy 1996; Tomasello et al. 1999), monkeys (Burkart and Heschl 2006; Emery et al. 1997; Tomasello et al. 1998) lemurs (Shepherd and Platt 2008), goats (Kaminski et al. 2005) and corvids. The most parsimonious explanation for gaze following into distant space is predator avoidance, which would also explain why such a broad range of social species appear to possess these skills. In line with this interpretation is our finding that corvids start to show this behaviour as soon as they become vulnerable to aerial predators; also, ravens are equally likely to follow siblings' and non-siblings' gaze, although they otherwise attend more readily to individuals they share socio-positive relationships with (Scheid et al. 2007). Whereas such a distinction may be useful when learning where to feed, it may be a deadly mistake when trying to avoid predators.

Our results cannot solve the question, if gaze following in mammals and birds is a result of convergent evolution or phylogenetic continuity. To close this gap, data on gaze following in social reptiles, i.e. iguanas or turtles, are needed. Furthermore, we lack data on the prevalence of the different modes of gaze following in vertebrates. Reports on geometrical gaze following abilities are still limited to apes (Bräuer et al. 2005; Povinelli and Eddy 1996; Tomasello et al. 1999) and corvids, with the only negative results so far stemming from jackdaws. The notion that geometrical gaze following appears to be restricted to few species led us to propose that geometrical gaze following may serve other functions than gaze following into distant space. Frequent caching out of sight of conspecifics and geometrical gaze following emerge at the same time during ontogeny, and these two skills may be linked functionally (Schloegl et al. 2007). An understanding of the functionality of barriers may allow to cache out of sight of potential thieves. Our findings from the food-caching rooks support this idea: geometrical gaze following emerges once more later in life, presumably when juvenile rooks became independent from their parents. In the non-caching jackdaw, we did not find evidence for geometrical gaze following skills. Thus far this idea is based on data from three closely related species only, all of which belong to the youngest genus within the corvids, *Corvus*. The common ancestor of all corvids was most likely a moderate food-cacher (de Kort and Clayton 2006). Therefore, jackdaws would have lost their gaze following skills, along with their caching skills, secondarily. Clearly, more comparative data on food-cachers from other corvid genera (e.g. nutcrackers, jays or magpies) are needed. In particular, it would be enlightening to learn more about the gaze following capabilities of the white-throated magpie jay (*Calocitta formosa*), the only other corvid species known to lack food caching behaviour entirely (de Kort and Clayton 2006). Alternatively, geometrical gaze following skills may have evolved within the corvids only after the phylogenetic branches of modern jackdaws and crows, consisting of all extent crow and raven species, had split between seven and ten million years ago (R. Fleischer, pers. comm.).

We do not propose that the caching of food *per se* is the key for developing advanced gaze following in corvids, but the necessity of securing caches against pilferers. Parids (chickadees and tits) also cache food, but appear to lack observational spatial memory (Baker et al. 1988), i.e. pilferer do not remember where they have seen others cache. In turn, food cachers may not need to cache out of sight to protect their caches against theft and geometrical gaze following skills may not be crucial for parids. Still, kangaroo rats (*Dipodomys merriami*; Preston and Jacobs 2005), grey squirrels (*Sciurus carolinensis*; Leaver et al. 2007) as well as some parids (Hitchcock and Sherry 1995; Stone and Baker 1989) have been reported to adapt their caching behaviour to the presence of competitors. Again, it is not known whether pilferer in these species need to observe a caching event for successful pilferage, but if food cachers in these species can assess whether others can see them while caching, we predict that they also possess geometrical gaze following skills.

To sum up, corvids are another model group for studying advanced gaze following abilities in non-human animals. Corvids seem to be comparable to great apes in their geometrical gaze following skills. Additionally, experimental procedures are available to test animals with human and conspecific models, and we have outlined a theoretical framework to promote comparative research on gaze following in non-human animals. Although knowledge about corvid gaze following is not sufficient to understand the mechanisms leading to the evolution of such skills, it may provide an important piece to the puzzle.

ACKNOWLEDGEMENTS

The Herzog v. Cumberland Game Park and the "Verein der Förderer KLF" provided permanent support. This work was funded by FWF projects R31-B03 and P16939-B03 and C. Schloegl was supported by a DAAD-Doktorandenstipendium. C. Scheid was supported by a studentship by the French Ministry of Research. Matthias-Claudio Loretto, Mareike Stöwe, Larisa Lee Cruz, Drew Gaede, Ruth Swoboda, Christine Schwab, Aude Erbrech and Thomas Hindelang provided help with the experiments. We are grateful for valuable comments on earlier versions of the manuscript by I. Scheiber, B. Heinrich and R. Noë. C. Schloegl is grateful to Dietrich von Holst for his support. We thank Paul Sömmer for his invaluable help while obtaining ravens from the wild and H.-U. Stuiber for his help while obtaining jackdaw chicks. The zoos in Wuppertal and München provided raven chicks.

REFERENCES

Anderson, J. R. and Mitchell, R. W. (1999). Macaques but not lemurs co-orient visually with humans. *Folia Primatologica 70*: 17-22.

Baker, M. C., Stone, E. R., E., B. A., Shelden, R. J. and Mantych, M. D. (1988). Evidence against observational learning in storage and recovery of seeds by black-capped chickadees. *The Auk 105*: 492-497.

Baron-Cohen, S. (1995). *Mindblindness: An Essay on Autism and Theory of Mind.* Cambridge, Massachusetts: MIT Press.

Bräuer, J., Call, J. and Tomasello, M. (2004). Visual perspective taking in dogs (*Canis familiaris*) in the presence of barriers. *Applied Animal Behaviour Science 88*: 299-317.

Bräuer, J., Call, J. and Tomasello, M. (2005). All Great Ape Species Follow Gaze to Distant Locations and Around Barriers. *Journal of Comparative Psychology 119(2)*: 145-154.

Brooks, R. and Meltzoff, A. N. (2005). The development of gaze following and its relation to language. *Developmental Science 8(6)*: 535-543.

Bugnyar, T. and Heinrich, B. (2005). Ravens, *Corvus corax*, differentiate between knowledgeable and ignorant competitors. *Proceedings of the Royal Society of London, Series B 272*: 1641-1646.

Bugnyar, T. and Heinrich, B. (2006). Pilfering ravens, *Corvus corax*, adjust their behaviour to social context and identity of competitors. *Animal Cognition 9*: 369-376.

Bugnyar, T., Stöwe, M. and Heinrich, B. (2004). Ravens, *Corvus corax*, follow gaze direction of humans around obstacles. *Proceedings of the Royal Society of London, Series B 271*: 1331-1336.

Bugnyar, T., Stöwe, M. and Heinrich, B. (2007). The ontogeny of caching in ravens, *Corvus corax*. *Animal Behaviour 74(4)*: 757-767.

Burkart, J. and Heschl, A. (2006). Geometrical gaze following in common marmosets (*Callithrix jacchus*). *Journal of Comparative Psychology 120(2)*: 120-130.

Butler, S. C., Caron, A. J. and Brooks, R. (2000). Infant understanding of the referential nature of looking. *Journal of Cognition and Development 1(4)*: 359-377.

Butterworth, G. and Jarrett, N. (1991). What minds have in common is space: spatial mechanisms serving joint visual attention in infancy. *British Journal of Developmental Psychology 9*: 55-72.

Byrne, R. W. and Whiten, A. (1988). *Machiavellian Intelligence: Social Expertise and the Evolution of Intellect in Monkeys, Apes and Humans*. Oxford: Clarendon Press.

Corkum, V. and Moore, C. (1995). Development of joint visual attention in infants. C. Moore and P. J. Dumham *Joint attention*. Hillsdale, NJ: Lowrence Earlbum Associates: 61-84.

Dawkins, M. S. (2002). What are birds looking at? Head movements and eye use in chickens. *Animal Behaviour 63*: 991-998.

de Kort, S. R. and Clayton, N. S. (2006). An evolutionary perspective on caching by corvids. *Proceedings of the National Academy of Science, USA 273*: 417-423.

Emery, N. J. (2000). The eyes have it: The neuroethology, function and evolution of social gaze. *Neuroscience & Biobehavioral Reviews 24*: 581-604.

Emery, N. J. (2006). Cognitive Ornithology: the evolution of avian intelligence. *Philosophical Transactions of the Royal Society of London, Series B 361(1465)*: 23-43.

Emery, N. J. and Clayton, N. S. (2004). The Mentality of Crows: Convergent Evolution of Intelligence in Corvids and Apes. *Science 306*: 1903-1907.

Emery, N. J., Lorincz, E. N., Perret, D. I., Oram, M. W. and Baker, C. I. (1997). Gaze following and joint attention in rhesus monkeys (*Macaca mulatta*). *Journal of Comparative Psychology 111*: 286-293.

Flavell, J. H., Everett, B. A., Croft, K. and Flavell, E. R. (1981). Young Children's Knowledge About Visual Perception: Further Evidence for the Level 1 - Level 2 Distinction. *Developmental Psychology 17(1)*: 99-103.

Gomez, J.-C. (2005). Species comparative studies and cognitive development. *TRENDS in Cognitive Science 9(3)*: 118-125.

Haffer, J. and Kirchner, H. (1993). *Corvus corax* - Kolkrabe. U. N. Glutz von Blotzheim and K. M. Bauer *Handbuch der Vögel Mitteleuropas*. Wiesbaden: AULA-Verlag. *13/III Passeriformes (4. Teil)*: 1947-2022.

Heinrich, B. (1989). *Ravens in winter*. New York: Summit Books of Simon & Schuster.

Heinrich, B. (1999). *Mind of the raven*. New York: Harper.

Heinrich, B., Kaye, D., Knight, T. and Schaumburg, K. (1994). Dispersal and association among common ravens. *The Condor 96*: 545-551.

Heinrich, B. and Pepper, J. W. (1998). Influence of competitors on caching behaviour in the common raven, *Corvus corax*. *Animal Behaviour 56*: 1083-1090.

Hitchcock, C. L. and Sherry, D. F. (1995). Cache pilfering and its prevention in pairs of black-capped chickadees. *Journal of Avian Biology 26*: 181-192.

Itakura, S. (1996). An exploratory study of gaze monitoring in nonhuman primates. *Japanese Psychological Research 38*: 174-180.

Itakura, S. (2004). Gaze-following and joint visual attention in nonhuman animals. *Japanese Psychological Research 46(3)*: 216-226.

Iwaniuk, A. N. and Hurd, P. L. (2005). The Evolution of Cerebrotypes in Birds. *Brain, Behavior and Evolution 65*: 215-230.

Kaminski, J., Riedel, J., Call, J. and Tomasello, M. (2005). Domestic goats, *Capra hircus*, follow gaze direction and use social cues in an object choice task. *Animal Behaviour 69*: 11-18.

Leaver, L. A., Hopewell, L., Caldwell, C. and Mallarky, L. (2007). Audience effects on food caching in grey squirrels (*Sciurus carolinensis*): evidence for pilferage avoidance strategies. *Animal Cognition 10*: 23-27.

Madge, S. and Burn, H. (1994). *Crows and Jays: A guide to the crows, jays and magpies of the world* Boston, MA: Houghton Mifflin.

Moll, H. and Tomasello, M. (2004). 12- and 18-month-old infants follow gaze to spaces behind barriers. *Developmental Science 7(1)*: F1-F9.

Povinelli, D. J. and Eddy, T. J. (1996). Chimpanzees: joint visual attention. *Psychological Science 7*: 129-135.

Premack, D. and Woodruff, G. (1978). Does the chimpanzee have a theory of mind? *Behavioral and Brain Sciences 4*: 515-526.

Preston, S. D. and Jacobs, L. F. (2005). Cache Decision Making: The Effects of Competition on Cache Decisions in Merriam's Kangaroo Rat (*Dipodomys merriami*). *Journal of Comparative Psychology 119(2)*: 187-196.

Scheid, C., Range, F. and Bugnyar, T. (2007). When, what and whom to watch? Quantifying attention in ravens and jackdaws. *Journal of Comparative Psychology 121(4)*: 380-386.

Schloegl, C., Kotrschal, K. and Bugnyar, T. (2007). Gaze following in Common Ravens (*Corvus corax*): Ontogeny and Habituation. *Animal Behaviour 74(4)*: 769-778.

Schloegl, C., Kotrschal, K. and Bugnyar, T. (2008a). Do common ravens (*Corvus corax*) rely on human or conspecific gaze cues to detect hidden food? *Animal Cognition in press*.

Schloegl, C., Kotrschal, K. and Bugnyar, T. (2008b). Modifying the object-choice task: Is the way you look important for ravens? *Behavioural Processes 77(1)*: 61-65.

Schwab, C., Bugnyar, T., Schloegl, C. and Kotrschal, K. (2008). Enhanced social learning between siblings in common ravens, *Corvus corax*. *Animal Behaviour in press*.

Shepherd, S. V. and Platt, M. L. (2008). Spontaneous social orienting and gaze following in ringtailed lemurs (*Lemur catta*). *Animal Cognition 11*: 13-20.

Stone, E. R. and Baker, M. C. (1989). The effects of conspecifics on food caching by black-capped chickadees. *The Condor 91*: 886-890.

Tomasello, M., Call, J. and Hare, B. (1998). Five primate species follow the visual gaze of conspecifics. *Animal Behaviour 55*: 1063-1069.

Tomasello, M., Hare, B. and Agnetta, B. (1999). Chimpanzees, *Pan troglodytes*, follow gaze direction geometrically. *Animal Behaviour 58*: 769-777.

Reviewed by: Bernd Heinrich

In: Animal Behavior: New Research   ISBN: 978-1-60456-782-3
Editors: E. A. Weber, L. H. Krause   © 2008 Nova Science Publishers, Inc.

Chapter 5

HOW DIFFERENT HOST SPECIES INFLUENCE PARASITISM PATTERNS AND LARVAL COMPETITION OF ACOUSTICALLY-ORIENTING PARASITOID FLIES (TACHINIDAE: ORMIINI)

Gerlind U. C. Lehmann[*]

Freie Universität Berlin, Institut für Biologie, Berlin, Germany

ABSTRACT

Sexual signals are often critical for mate attraction and reproduction but their conspicuousness can expose the signallers to parasites and predators. In orthopteran insects males typically produce acoustic signals to attract females for mating. Tachinid flies of the family Ormiini act as illicit receivers by detecting the mating songs of their bushcricket and gryllid hosts. Ormiini flies can be characterised as opportunistic hunters; the taxonomic specificity of these flies is moderately low and they can have a range of alternative hosts.

This chapter is in two parts. In the first I quantify the influence host species has on body size and life history traits of different populations of *Therobia leonidei*, the only European Ormiini fly species. For parasitoids host size can constrain offspring growth, subsequently influencing the evolution of body size and life history traits. I compared fly populations developing in two hosts: *Poecilimon mariannae* and in the lighter *P. thessalicus*. The fly populations investigated had no substantial morphological or molecular-genetic differences. The general pattern of larval competition was similar in both hosts; increasing parasitoid brood size reduced pupal weight and survival to adulthood. Consistent with a local adaptation hypothesis, pupal weight in the heavier host was about 30% heavier than are those parasitizing the lighter host. Similarly, the critical

---

[*] PD Dr. Gerlind U.C. Lehmann. Freie Universität Berlin, Institut für Biologie, Abteilung Evolutionsbiologie, Königin-Luise-Straße 1-3, 14195 Berlin, Germany. E-mail: gerlind.lehmann@t-online.de; *www.guclehmann.de*

weight necessary for successful hatching of fly maggots was significantly lower in the lighter host. In contrast, brood size was similar between host species.

In the second part I review studies of gryllid and tettigoniid parasitizing Ormiini flies. The general pattern of host usage for temperate species is quite similar, with pupae from single infected hosts weighing around 10 percent of host weight. One remarkable exception is the Australian *Homotrixa alleni*, a very large fly that parasitizes extremely heavy bushcricket species. In this fly species single pupa weigh less than four percent of host weight, leading to higher mean parasitoid clutch sizes. Coupled to this reduced parasitoid-to-host weight ratio is prolonged host survival and an increase in the probability of superparasitism.

Keywords: Clutch size, body size, host size, host-parasitoid relationship, *Therobia*, Ormiini, Tachinidae, *Poecilimon*, Orthoptera

INTRODUCTION

Insect parasitoids are an active field of ecological and evolutionary research due to the recognition that host-parasitoid systems offer unparalleled opportunities to examine fundamental questions in animal behaviour and evolutionary ecology (Eggleton & Belshaw 1992). However, most research has focused on Hymenoptera parasitoids (Godfray 1994, Hassell 2000, Hochberg & Ives 2000). These parasitoids have a relatively limited range of hosts and the parasitoid habit appears to have evolved only once in the Hymenoptera (Eggleton & Belshaw 1993). In contrast Dipteran parasitoids are exceedingly diverse in both their hosts and their evolutionary origins (Feener & Brown 1997). Tachinidae is one of the most diverse and ecologically important parasitoid families in the order Diptera. As parasitoids they are important natural enemies in most terrestrial ecological communities. One of the few traits that unites this diverse assemblage of flies is that all tachinids are parasitoids of arthropods, especially insects. Because of their predominance as parasitoids of the larval stage of Lepidoptera and other major groups of insect herbivores, tachinids often play significant roles in regulating herbivore populations, with over 100 species employed in biological control programs of crop and forest pests (Stireman et al. 2006). Despite their diversity and ecological impact relatively little is known about the basic ecology of tachinids.

Variation among hosts is an important determinant of phenotypic variation in insects that use discrete resources, such as parasitoids (Godfray 1994). Host size is a major source of phenotypic variation as it may constrain the evolution of body size and life history traits (Hardy et al. 1992, Allen and Hunt 2001, Mackauer and Chau 2001, Häckermann et al. 2007). One of the major decisions gregarious parasitoids have to make is how many eggs or larvae to lay on or in a host (Godfray et al. 1991). Host size is often considered to be a key factor influencing parasitoid fitness (Godfray 1994), so female parasitoids are expected to lay the number of eggs that maximizes their gain in fitness. The resulting intraspecific larval competition from gregarious oviposition influences many aspects of adult fitness, especially when host size is fixed. Intraspecific competition occurs when two or more individuals of the same species strive for the same resource. Intraspecific competition can be in the form of a scramble (resources divided equally) or a contest (resources divided unequally). Under contest competition there are winners and losers. Scramble competition arises when a resource is equally partitioned among competing larvae, with all competitors suffering the consequences of limitation equally. In species with scramble competition, individuals in populations adapted to large hosts are generally larger than those adapted to small hosts.

Several theoretical and empirical studies have focussed on clutch-size patterns in hymenoptera parasitoids (summarized in Godfray 1994). In contrast, dipteran parasitoids have received less attention (Feener & Brown 1997). Tachinids differ from Hymenoptera in lacking the rigid ovipositor that provides information about host parasite status and quality, and in often depositing mobile planidia larvae rather than eggs (Stireman et al. 2006). These differences may lead to female Tachinids not being able to reliably assess the number of larvae entering a given host (Adamo et al. 1995b).

Research on the small tribe Ormiini within the Tachinidae was undertaken by evolutionary biologists (Cade 1975, Zuk & Kolluru 1998) who realized that their unique exploitation of acoustic mating songs for finding hosts may impose strong natural selection on signalling (Zuk et al. 1993, 1995), signal structures (Rotenberry et al. 1996; Zuk et al. 1998, 2001; Lehmann & Heller 1998) and even the evolution of alternative mating strategies (Cade 1975, 1981; Zuk et al. 2006). Tachinid flies of this tribe rely on acoustic cues to locate hosts (Cade 1975, Walker 1986, Gray et al. 2007) and their ear is tuned towards the frequency of host's song (Lakes-Harlan and Heller 1992, Robert et al. 1992, Stumpner et al. 2007). Deposition of live planidia (larviposition) is elicited by host calling (Allen et al. 1999). After landing, female Ormiini flies deposit planidia either directly on the host or spread a number of larvae in their vicinity (Adamo et al. 1995b; Cade 1975), where they are able to survive for up to two hours (Allen et al. 1999). In one study more than fifty percent of flies had no direct contact with the host, but laid a number of planidia nearby (Allen et al. 1999). Planidia penetrate the host's interscleral membranes and a few days later produce a connection with the air outside through a funnel in the host's abdomen (Leonide 1969). Infected hosts inevitably die following parasitoid emergence.

The objective of the first part of this review is to quantify the relative contribution of environmental (host species) effects to differences in body size and life history traits between populations of the Ormiini *Therobia leonidei* MESNIL, 1964 developing on lighter and heavier host species. This European fly species has been found to attack the singing sex of three different families of bushcrickets (Leonide 1969, review in Lehmann 2003). Infection by planaria is indicated by their breathing funnel which can be seen as a brown dot on the ventral side of the male's abdomen, normally four days after infection (Lehmann and Heller 1998, Lehmann 2006). The average survival time following the appearance of the breathing funnel is seven days. The longest time that parasitized males have survived is 14 days after fly attack (Lehmann and Lehmann 2006). Developing fly larvae reduce male bushcricket survival and reproductive effort. Parasitism reduces a male host's investment capability into spermatophore production (Lehmann and Lehmann 2000a) which results in a shorter female mating refractory period and fewer eggs being laid (Lehmann and Lehmann 2000b). Infected males call less and are discriminated against by females (Lehmann and Lehmann 2006). Although *T. leonidei* is considered a generalist because of the number of known hosts, host use varies substantially between localities, and most populations attack single or a few hosts and are thus selective relative to the diversity of bushcricket species available to them (Lehmann 2003). This may lead to adaptation to local hosts (Fox & Czesak 2000).

In populations of the Greek phaneropterine bushcricket *Poecilimon mariannae* WILLEMSE AND HELLER, 1992 males were found to be parasitized in large proportions, up to

65% at the end of their breeding season (Lehmann & Heller 1998). Mute females were never found to be infected (Lakes-Harlan and Heller 1992, Lehmann and Heller 1998). I was able to study the parasitism behaviour in a further, closely related host species *P. thessalicus* BRUNNER VON WATTENWYL, 1891 (Lehmann et al. 2001). Although this species showed a remarkable variation in body size between populations (Lehmann and Lehmann in press.), all body sizes are smaller than in *P. mariannae*. Larger hosts are expected to be more advantageous in terms of offspring fitness than smaller hosts because they contain a greater quantity of resources. In gregarious species, fitness is affected not only by host size but also by the number of parasitoids developing in a single host for hymenoptera (Zaviezo & Mills 2000, Bell et al. 2005, Häckermann et al. 2007) and tachinid parasitoids (Nakamura 1995, Reitz & Adler 1995, Allen & Hunt 2001, Kolluru & Zuk 2001). Therefore, I assessed the influence of host species differing in body size on the success of the parasitoid female in terms of brood size and number of surviving offspring.

Part II

What is known about Ormiini tends to be widely dispersed in specialized articles and even unpublished reports. In the second part of this chapter I review Ormiini ecology, focusing on parasitism patterns and host usage. What we know comes largely from four Ormiini-host systems: my own studies on *Therobia leonidei,* which attacks several bushcricket hosts in Europe, the Australian *Homotrixa alleni* which is a parasitoid of the large weight bushcricket *Sciarasaga quadrata* (work of Geoff Allen and collaborators), and two species of the genus *Ormia* which infects crickets or mole crickets (see especially the work of Adamo, Cade, Fowler, Frank, Kolluru, Walker and Zuk).

I review four main areas: (*a*) life history data, (*b*) parasitism rates and seasonality of occurrence, (*c*) evolution of reproductive strategies including brood sizes and superparasitism. Superparasitism is defined as the deposition of a clutch of eggs by a female parasitoid in or on a host already parasitized by itself or a conspecific female (van Alphen & Visser 1990, Godfray 1994), and (*d*) oviposition strategies.

The evolution of acoustic hunting, biogeography and host ranges in Ormiini have been reviewed previously (Lehmann 2003).

MATERIALS AND METHODS

Part I

Parasitoid Flies

Therobia leonidei is a parasitoid fly (Diptera: Tachinidae: Ormiini) distributed in Southern Europe in the Mediterranean Countries. The northernmost record is from the Tessin (Switzerland), in the East it is recorded from Turkey, the Ukraine and Azerbaidzhan (Lehmann 1998 and references therein).

After locating a male bushcricket, the female fly deposits planidia larvae on the abdomen of the male, which burrow inside the host. A few days later, the larvae produce a connection with the air outside the host through a 'breathing funnel' in the host abdomen (Leonide 1969). Subsequently, the infection by the fly is indicated by a brown dot on the ventral side of the male's abdomen (Lehmann 2006), in *Poecilimon mariannae* after four days (Lehmann & Heller 1998).

The Bushcricket Host Species

Therobia leonidei was found to parasitize 13 host species from three different bushcricket subfamilies (Lehmann 2003). After screening a large number of populations for parasitoid occurrence in Greece (see Lehmann 1998, Lehmann & Heller 1998), we analyzed parasitism patterns and host usage in two species of the phaneropterine genus *Poecilimon*. The bushcrickets *Poecilimon mariannae* WILLEMSE ET HELLER, 1992 and *P. thessalicus* BRUNNER VON WATTENWYL, 1891 are medium-sized species (mean body length around 2 cm), morphologically and ecologically quite similar and are members of the *Poecilimon propinquus*-group, which contains eight species (Lehmann AW 1998, Lehmann AW et al. 2006). I compared fly parasitism adapted to either the heavier bushcricket host *P. mariannae* (average weight of 560 mg) with a population adapted to the much lighter host *P. thessalicus* (average weight of 500 mg). The first host species *P. mariannae* is confined to areas of central Greece and inhabits roadside verges and grassy patches (Willemse & Heller 1992). The population studied was on a grassy ridge near Vrissiá (22°19' E, 39°15° N, Nomos Lárissa; see Lehmann and Heller 1998). The second and lighter host species was examined in Elatohori (Central Greece, Pieria-Mountains, 22°16 E, 40°19 N), where it is found on shrubs and bushes (Lehmann et al. 2001). There is considerable variation in body size between *P. thessalicus* populations (Lehmann & Lehmann in press). However the Elatohori population was one of the few parasitized populations of this species (Lehmann unpubl. data).

Parasitism Patterns

Field samples of adult male bushcrickets were screened for the presence of breathing funnels as an indicator of parasitoid infection. The breathing funnel appears four days after infection (Lehmann & Heller 1998). To measure levels of parasitism the populations of *P. mariannae* at Vrissia and *P. thessalicus* at Elatohori were investigated over three consecutive years 1994-1996. The rate of parasitism was observed over the whole season, from the beginning of male calling until the fading of male populations (see Lehmann and Lehmann 2006 for a life table analysis of the larger host species *P. mariannae*).

In 1995 all individuals were not only checked for presence or absence of parasitoid breathing funnels, but additionally the number of maggots in a male was counted via the number of breathing funnels. To verify that the number of fly larvae corresponded to the number of observed breathing funnels I dissected twenty males from each species. The number of visible funnels was always identical with the number of maggots found inside the males.

As it was not possible to determine whether all of the larvae inside a host were the progeny of a single parasitism event I refer to brood size and avoid the term clutch size. Furthermore, clutch size, defined as the number of planidia placed in a single event, might be larger than brood size because not all planidia might be successful in entering a host. After emergence and pupating, all pupae were weighed (wet weight) and hatched flies were sexed. For comparison host males were weighed on a Kern EG 300-3M scale (± 1 mg).

RESULTS

Parasitism in two closely related bushcrickets infected by *Therobia leonidei* in Greece.

Parasitism Patterns

Male parasitism in *P. mariannae* reached 57, 50 and 54 percent in three consecutive years (Figure 1a). Rates of parasitism increased nearly linear as the calling season progressed. While there is variation in parasitism rate in the middle of the calling season the general pattern is similar between years.

In the second host species, *P. thessalicus,* substantial variation was observed in the parasitism pattern between three consecutive summers (Figure 1b). In 1994 the level of parasitism was low and did not exceed more than 10 percent during the whole season. In 1996 parasitism was slightly higher than the previous two years but showed a similar pattern to 1994 with a peak parasitism rate in the middle of the calling season. In contrast, parasitism in 1995 showed a linear increase reaching a plateau of 40 percent infection by the end of the season.

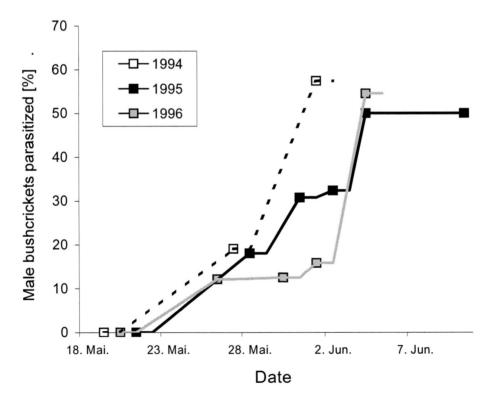

Figure 1a. Percent parasitism of male *Poecilimon mariannae* by *Therobia leonidei* in the years 1994-1996. The beginning of the time - axis corresponds with the moulting of the first males to adulthood. The very last data represent the last sight of adult males. Sample sizes were *1994:* n = 50, 21, 54 - *1995:* 50, 50, 13, 34, 36, 12 - *1996:* 50, 33, 8, 57, 33.

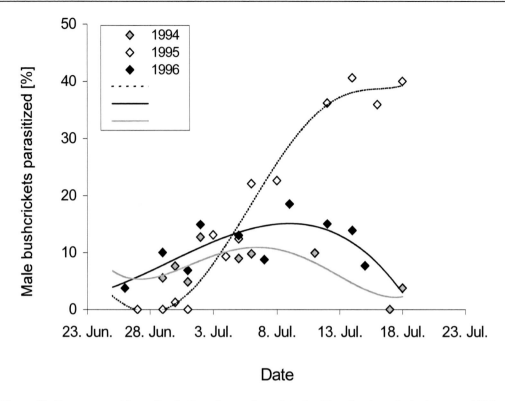

Figure 1b. Percent parasitism of male *Poecilimon thessalicus* by *Therobia leonidei* in the years 1994-1996. The beginning of the time - axis corresponds with the moulting of the first males to adulthood. The very last data represents the last sight of adult males. Sample sizes were *1994*: n = 54, 118, 41, 78, 82, 111, 58, 53 - *1995*: 65, 17, 80, 64, 130, 161, 113, 77, 84, 105, 32, 64, 25 - *1996*: 53, 100, 58, 94, 100, 80, 81, 60, 65, 39.

Over the whole calling season of 1995, more than 50 percent of *P. mariannae* males had only one fly maggot (you need to be consistent and use only "larva" or "maggot"), a further 22 percent harboured two fly larvae, while hosts with three or more maggots were rarely found (Figure 2). As a maximum, seven breathing funnels were observed in individuals in 1995, although I have found a male *P. mariannae* in another population with twelve breathing funnels (unpubl. obs.). In the smaller host *P. thessalicus* fewer males had solitary fly larvae (38 percent) but the incidence of three or more parasitoids was nearly identical between host species.

While the proportion of parasitized males increased over the host's calling season, the mean brood size showed no trend over the season (Figure 3). Mean clutch size was statistically independent of time for both males of *P. mariannae* (Kruskal-Wallis-Test: H = 15.5, df = 13, p = 0.28) and *P. thessalicus* (Kruskal-Wallis-Test: H = 9.96, df = 8, p = 0.27). Over the whole season, parasitoid brood size (mean ± SE) in the heavier host species *P. mariannae* (2.13 ± 0.18) was marginally higher than in the lighter host *P. thessalicus* (1.92 ± 0.09), but this difference was not significant (Mann Whitney-U-test: Z = -0.15, p = 0.88).

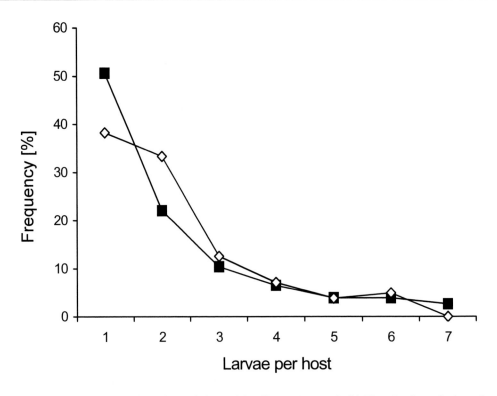

Figure 2. Frequency distribution of brood sizes of the dipteran's parasitoid *Therobia leonidei* in male bushcrickets of *Poecilimon mariannae* (■, n = 77) and *P. thessalicus* (◇, n = 183).

Brood size in *P. thessalicus* was to some extent variable over the season (Figure 4) and brood sizes larger than three may represent superparasitism (Figure 4 hatched bars). For the whole sampling season its amount was 8.6 percent, with a maximum of 17.6 percent. However, proposed superparasitism occurred in the middle and at the end of hosts calling season, when daily attack rates were highest (indicated by the strongest daily increase in absolute parasitism rates, see Figure 1b).

Nonetheless, superparasitism was obviously disadvantageous, as never more than one maggot hatched from *P. thessalicus*, regardless of brood size. In the larger host *P. mariannae*, two or even three maggots occasionally emerged from parasitized hosts. In no case involving brood sizes greater than one did I find all larvae surviving to pupation. For *P. mariannae* hosts where the brood size was equal or larger than four (proposed to be superparasitism) never more than a single parasitoid hatched.

Larval Competition

There is no correlational evidence that higher host weight resulted in higher parasitoid weight within a species. Parasitoid pupae hatched from single infected hosts showed no increase in weight with increasing host weight (Figure 5). A linear regression accounted for two ($y_{mariannae.}= 47.88 + 0.01$ x, $R^2 = 0.02$, n = 33) or seven percent of the total variation ($y_{thessalicus.}= 26.51 + 0.03$ x, $R^2 = 0.07$, n = 27). Bushcricket males of *P. mariannae*

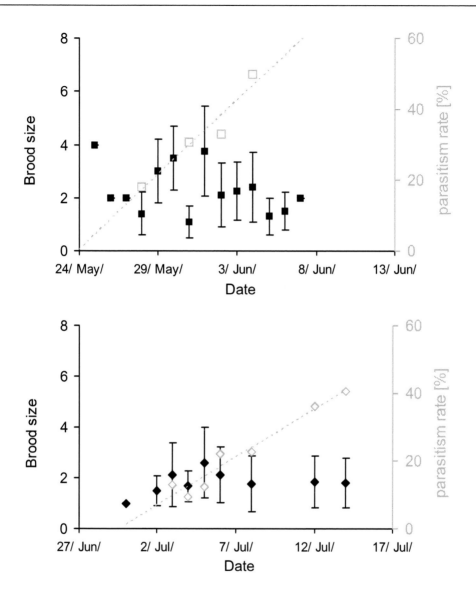

Figure 3. Brood size (mean ± SD) of *Therobia leonidei* over a period of two weeks in males of *Poecilimon mariannae* (top) and *P. thessalicus* (bottom). The parasitism rate of the whole populations in 1995 and the regressions line were plotted for comparison (see Figure 1). Sample sizes *P. mariannae*: n = 1, 1, 1, 10, 3, 2, 9, 3, 16, 3, 22, 3, 2, 1- *P. thessalicus*: 1, 4, 17, 15, 15, 17, 37, 45, 11.

were significantly heavier (563.91 ± 69.33 mg; mean ± SD) than *P. thessalicus* males (502.78 ± 78.84 mg) (t-test: T = 3.19, df = 58, p = 0.0023) at infection. The pupae from the relatively heavier host *P. mariannae* were even relatively heavier in relation to host body weight (9.59 vs. 8.33 %).

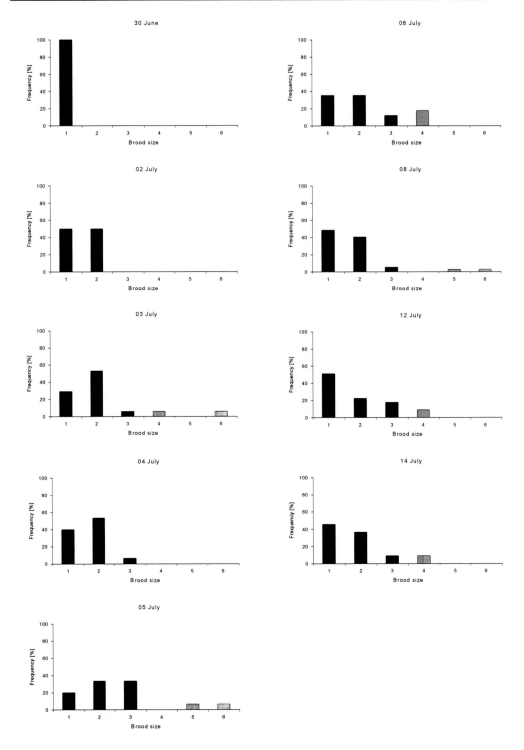

Figure 4. Seasonal progress of parasitoid brood size in *P. thessalicus* and the probable amount of superparasitism (hatched bars).

Parasitoid size was influenced by host size, resulting in substantially heavier pupae from *P. mariannae* compared to pupae from the smaller host species *P. thessalicus* (pupal weight: 53.39 ± 5.36 versus 41.30 ± 8.95 mg mean +- SD, two-way ANOVA $_{\text{host species}}$: $F_{1,47} = 32.73$, p = 0.000). The sex of the resulting fly had no significant influence on pupal weight (ANOVA $_{\text{fly sex}}$: $F_{1,47} = 1.21$, P = 0.28).

Bushcricket body weight did not predict the number of parasitoid larvae for either host (*P. mariannae* linear regression: $R^2 = 0.000$, n = 44, Figure 6top; *P. thessalicus* (regression: $R^2 = 0.003$, n = 104, Figure 6bottom).

Pupal weight declined significantly with increasing brood size in both host species (Figure 7) (two-way ANOVA $_{\text{clutch size}}$: $F_{4,175} = 17.15$, p = 0.000) with solitary pupae around 20 percent heavier than pupae from multiple-infected hosts. In all brood sizes pupae developing in *P. mariannae* were on average 32 to 44 percent heavier than those developing in *P. thessalicus* (two-way ANOVA $_{\text{host species}}$: $F_{1,175} = 86.44$, p = 0.000).

Pupal weight was a predictor for emergence success of adult flies (Figure 8). In pupae heavier than 35 mg hatching success was high in *P. mariannae* (≥ 80 %), whereas hatching success in *P. thessalicus* was much lower for any given weight. The minimum pupal weight result in a live fly was 27 mg in *P. mariannae* and less than half of that (11 mg) in *P. thessalicus*. No obvious differences between the sexes in both species could be found.

Pupal weight declined with brood size, and pupal weight was crucial for hatching success of adult flies. Therefore an increase in brood size resulted in a decline in the proportion of larvae that developed into adult flies in both host species (Figure 9). Fly larvae emerged from hosts more often if they had developed alone. Most parasitoids developed successfully from solitary infected male hosts into adult flies (73 % and 60 %), whereas in fivefold or more parasitized males only 13 and 10 percent successfully emerged as adults.

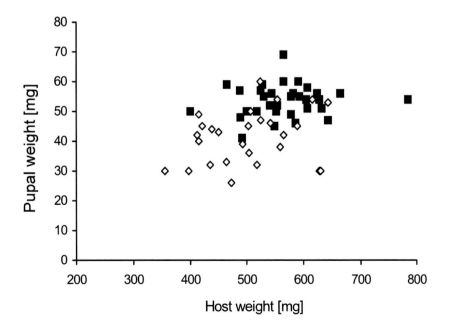

Figure 5. Pupal weight [mg] in relation to host weight [mg] from single infected host males of *Poecilimon mariannae* ($\blacksquare$, n = 33) and *P. thessalicus* ($\diamondsuit$, n = 27).

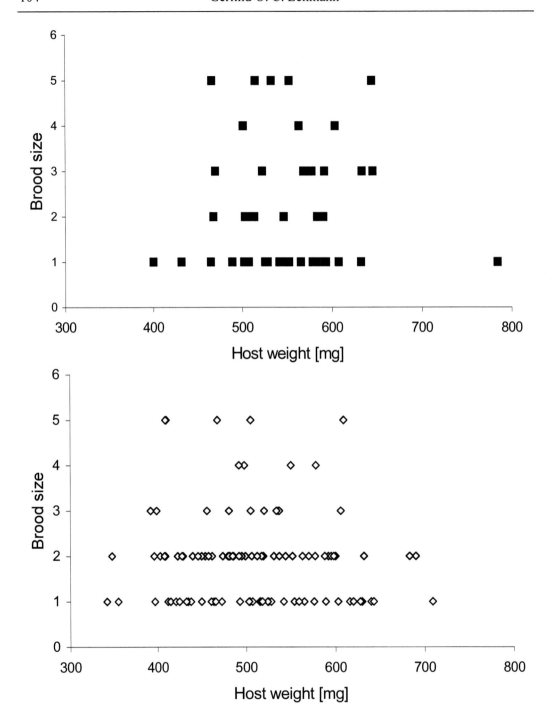

Figure 6. Brood size of *Therobia leonidei* in relation to host weight in *Poecilimon mariannae* (■, n = 44) and *P. thessalicus* (◇, n = 104).

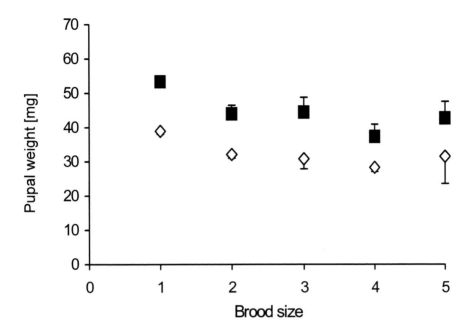

Figure 7. Mean (± SD) weight of pupae emerging from *Poecilimon mariannae* (■) or *P. thessalicus* (◇) males in relation to brood size. Sample sizes *P. mariannae*: n= 34, 23, 11, 9, 7 - *P. thessalicus*: 99, 91, 21, 13, 4.

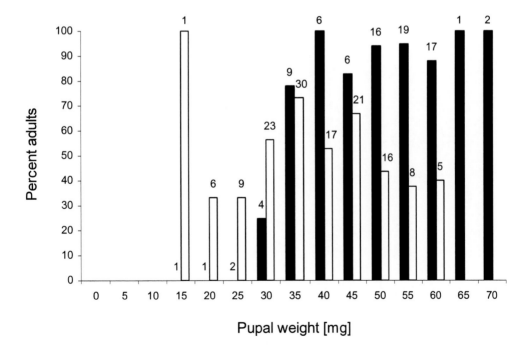

Figure 8. Hatching success of adult flies depending on pupal weight in *Poecilimon mariannae* (black bars, n = 84) and *P. thessalicus* (white bars, n = 136). Sample sizes presented above each bar.

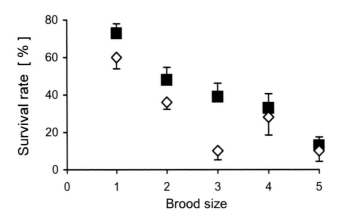

Figure 9. Mean (± SD) survival rate to adult flies from *Poecilimon mariannae* (■) or *P. thessalicus* (◇) males in relation to brood size. Sample sizes *P. mariannae*: n = 69, 26, 13, 10, 11 - *P. thessalicus*: 177, 93, 26, 9, 6.

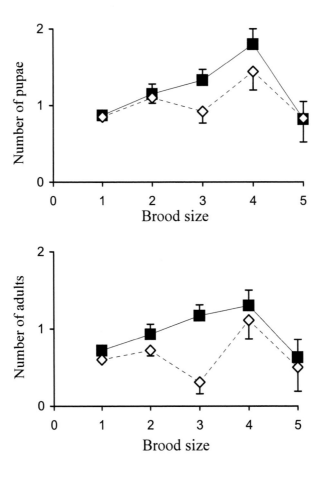

Figure 10. Mean (± SD) number of (top) pupae and (bottom) adult flies emerging from *Poecilimon mariannae* (■) or *P. thessalicus* (◇) males in relation to brood size. Sample sizes *P. mariannae*: n = 69, 26, 12, 10, 11 - *P. thessalicus*: 112, 62, 18, 8, 6.

The total number of hatched maggots (Figure 10top) increased up to a brood size of four maggots (two-way ANOVA$_{clutch\ size}$: $F_{4,266} = 8.56$, $p < 0.001$) independent of host species (ANOVA$_{host\ species}$: $F_{1,266} = 1.59$, $p = 0.21$). In accordance with the number of pupae, the number of adult flies (Figure 10bottom) nearly doubled in fourfold parasitized males compared to solitary infected ones for both host species (two-way ANOVA$_{clutch\ size}$: $F_{4,279} = 3.41$, $p = 0.0096$). A brood size larger than four reduced the number of adult flies to a value lower than for single infections. However, the number of adult flies was significantly influenced by host species (ANOVA$_{host\ species}$: $F_{1,279} = 7.44$, $p = 0.0068$), because a very reduced number of flies hatched from *P. thessalicus* males infected with three larvae.

For parasitoids from single infected hosts, the sex ratio of hatching flies was nearly equal, with 46.7 and 47 percent females for the two host species. A greater number of female flies emerged when brood size was larger than three in the heavier host *P. mariannae* or larger than two for the lighter host *P. thessalicus* (Figure 11). However, the increase in female parasitoid flies was not statistically significant for either species (*P. mariannae* $\chi^2 = 6.94$, df = 4, $p = 0.14$, n = 60; *P. thessalicus* $\chi^2 = 8.56$, df = 4, $p = 0.07$, n = 124), probably due to low sample sizes for higher brood sizes.

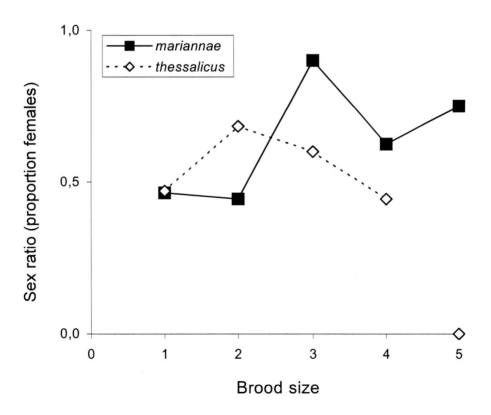

Figure 11. Sex ratio as an amount of females emerging from the heavier host *Poecilimon mariannae* (■) or the lighter host *P. thessalicus* (◇) in relation to brood size. Sample sizes *P. mariannae*: n = 28, 9, 10, 8, 4 - *P. thessalicus*: 66, 41, 5, 9, 3.

Host-parasitoid Aspects of Ormiini Flies - A Review

Life History

Even after three decades of active research starting with Cade (1975), we still have only limited knowledge on Ormiini life histories. The host ranges of over 80 percent of Ormiini species are not known, and the host range can only be approximated in four to five species. The majority parasitize singing bushcrickets, with the derived pattern of cricket parasitism occurring only in a subgroup of the genus *Ormia* (Lehmann 2003). Targeted hosts are likely to be singing males, which are active at night. Summaries of life history data for five Ormiini-Orthoptera systems are in Table 1.

Despite the different hosts used by *Therobia leonidei* and *Ormia ochracea* they have much in common: host weight, parasitoid pupal mass, relative pupal weight, and mean brood sizes are nearly equal. In contrast, the Australian *Homotrixa alleni* has a slower life-cycle, with a much longer developmental time inside the host and a prolonged pupal duration even if corrected for rearing temperatures. This might be directly related to the very large host species that they exploit, leading to a low relative pupal weight. The longer developmental time of the maggots might also give rise to the two-fold larger brood size, even if the number of maggots placed by a single fly female is similar to *Ormia ochracea*. This large brood size in *Homotrixa alleni* is best interpretated as resulting from superparasitism.

The direct pupal development found in all studies of Ormiini (Wineriter & Walker 1990, Allen 1995, Lehmann unpubl. data) is puzzling. Direct developing flies might eclose and be searching for hosts beyond the time singing males are present in the field, so we would expect an off-season pupal stage. We do not know whether this direct development is an artificial response to our rearing conditions or flies are using alternative hosts over the season. Experiments manipulating pupal duration in the field have been unsuccessful (Allen 1995, Lehmann unpubl. data).

Parasitism Rates and Seasonality

Ormiini flies are strictly nocturnal, most active shortly after sunset (Fowler 1987a; Cade et al. 1996; Allen 1998; Kolluru 1999). Due to the phonotactic host locating tactics of the flies, mostly singing males suffer from parasitism (Cade 1975, Lakes-Harlan & Heller 1992, Robert et al. 1992, review in Lehmann 2003). Parasitism patterns differ between the studied host-parasitoid systems according to seasonality and life-history of host populations. The greatest differences exist between crickets (including mole crickets) (Figure 12) and bushcrickets (Figure 13). Both groups differ fundamentally in their habitat use. Bushcrickets climb up and inside vegetation with males singing from relatively exposed positions. This might be an advantage for song propagation but in turn exposes singing males well to the sound locating parasitoid flies. In contrast crickets and mole crickets shelter inside self-constructed burrows. This habit might protect crickets to some extent from the parasitoids.

Table 1. Life history data of four Ormiini species

Fly species	Therobia leonidei	Ref.	Therobia leonidei	Ref.	Ormia ochracea	Ref.	Ormia depleta	Ref.	Homotrixa alleni	Ref.
Host species	Poecilimon marmarae		Poecilimon thessalicus		Teleogryllus oceanicus		Scapteriscus spp.		Sciarasaga quadrata	
Host Family	Tettigoniidae		Tettigoniidae		Gryllidae		Gryllotalpidae		Tettigoniidae	
	Bushcricket		Bushcricket		Cricket		Molecricket		Bushcricket	
Host weight [mg] (mean ± SD)	563.91 ± 69.33 (n=33)	(1)	502.78 ± 78.84 (n=77)	(1)	544.17 ± 133.51 (n=6)	(4)	range 540 - 1590	(8)	3540 ± 50 (n=35)	(12)
Pupal weight [mg] (mean ± SD)	53.39 ± 5.36 (n=33)	(1)	41.30 ± 8.95 (n=77)	(1)	47.86 ± 11.43 (n=14)	(4)	means around 52 mg	(8)	116 ± 5 (n=19)	(12)
Relative pupal weight (%)	9.47	(1)	8.21	(1)	8.80	(4)	(nearly 5 %)	(8)	3.28	(12)
Pupal to host weight	independent	(1)	independent	(1)			linear increase	(8)		
Brood size (mean ± SD)	2.13 ± 1.58 (n=77)	(1)	2.20 ± 1.38 (n=183)	(1)	1.77 ± 1.2 (n = 214)	(5)			3.48 ± 2.67 (n=114)	(12)
					1.73 ± 1.0 (n = 22)	(6)				
	range 1-7	(1)	range 1-6	(1)	ranges 1-5, 1-8	(5,6)	range 0-2	(9)	range 1-12	(12)
Brood size and season	independent	(1)	independent	(1)					increases	(12)
Brood size to host weight	independent	(1)	independent	(1)	independent	(5,6)			independent	(12)

Table 1. (Continued)

	Ref.		Ref.		Ref.		Ref.		Ref.
Fecundity of parasitoid (mean ± SD)				165 ± 192 (n=248)	(5)	175.5 ± 90.96 (n=111)	(10)	range 94-600 (n=53)	(13)
(=number of planidia)				range 0-430		range 28-488	(10)		
				219 ± 38 SE (n=11)	(7)	148.2 ± 90.2 (n=30)	(10)		
				range 65-517		range 32-400			
						range 36-486	(11)		
						187 ± 45 SE (n=6, lab)	(7)		
						range 70-310			
Parasitoid fecundity to body size				increases	(5)			increases	(13)
Planidia laid / field				5.2 ± 2.8 (n=8)	(6)				
				range 3-8					
Planidia laid / laboratory				6.1 ± 5.2 (n=8)	(6)			4.3 ± 0.7 (n=90)	(14)
				range 2-18				range 0-10	
Parasitoid inside host [d]		11	(3)	7-9 at 25°C	(7)	8-9 at 23-25°C	(7)	up to 17 days	(12)
Time until breathing funnel [d]		4	(2)						
Breathing funnel until death [d]		7	(3)					up to 13 days	(12)
Pupal developing at 25°C [d]		13 {at 17° / 28° C}		12-15	(7)	11-13	(7)	21 ± 0.7, 22 ± 0.2	(12)

Data are compiled from following references: (1) this book chapter part I, (2) Lehmann & Heller 1998, (3) Lehmann & Lehmann 2006, (4) Kolluru pers. comm., (5) Kolluru & Zuk 2001, (6) Adamo et al. 1995b, (7) Wineriter & Walker 1990, (8) Welch 2006, (9) Frank pers. comm., (10) Fowler 1987, (11) Fowler & García 1987, (12) Allen 1995, (13) Allen & Hunt 2001, (14) Allen et al. 1999.

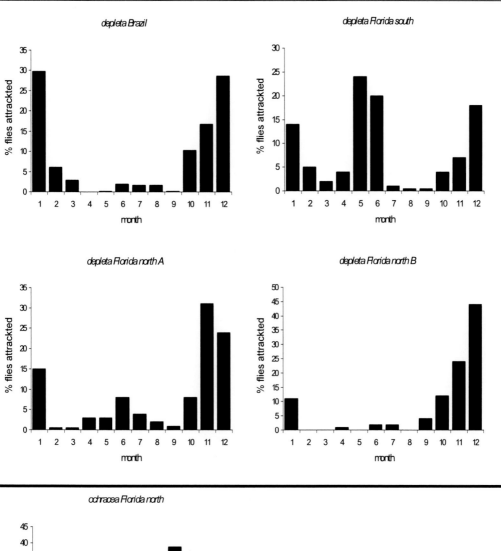

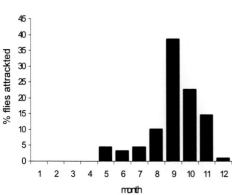

Figure 12. Seasonal captures of female parasitoid flies on sound traps using synthesized songs. Mole cricket songs were tested in Brazil (Fowler 1987b) and extensively in different localities in Florida (Walker et al. 1996), attracting females of *Ormia depleta*. Songs of the cricket *Gryllus rubens* (below the bold line) were also tested in Florida (Mangold 1978, see Walker 1986 for similar data) attracting *Ormia ochracea*. Fly numbers attracted and courted were normalized to percentage values over the whole season for comparison.

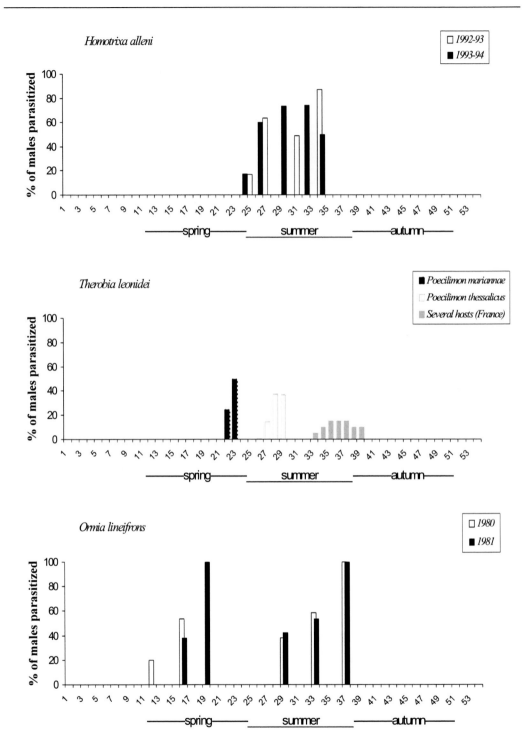

Figure 13. Percent parasitism in three different bushcricket parasitoids. On top the parasitism rates of the Australian Fly *Homotrixa alleni* in its main host *Sciarasaga quadrata* in two consecutive seasons (reanalyzed data from Allen 1995). In the middle the parasitism patterns of the European *Therobia leonidei* in different hosts (original data, Leonide 1969 for France). In northern Florida *Ormia lineifrons* is bivoltine utilizing the same host *Neoconocephalus triops* (reanalyzed data from Burk 1982).

It is not surprising, therefore, that bushcrickets are the hosts for the majority of Ormiini flies (Lehmann 2003) and that parasitism rates are quite high in all studied species, particularly at the end of the season (Figure 12). In Australia the bushcricket *Sciarasaga quadrata* is the main host of *Homotrixa alleni*, with parasitism occurring over the whole southern summer from November to February.

The European fly species *Therobia leonidei* has been observed parasitizing several bushcricket species. The main hosts in Greece are members of the phaneropterine genus *Poecilimon*, who have a calling season of up to four weeks (Lehmann & Lehmann 2006). *P. mariannae* is a lowland species whose breeding season commences earlier than the higher altitude *P. thessalicus*. However, in France the fly has been observed in late summer and autumn in other bushcricket species as well (Figure 12 middle). In North America the bushcricket *Neoconocephalus triops* is parasitized by *Ormia lineifrons*. Fly parasitism is bivoltine in northern Florida (Figure 12 bottom), but extends at least over four summer months in southern Florida (Burk 1982). No data currently exist for tropical bushcricket species.

In bushcrickets singing males are generally the sole host. There are a few exceptions such as where low numbers of acoustically responding females or even nymphs are found parasitized (reviewed in Lehmann 2003).

Reported parasitism rates are generally much lower in crickets than in bushcrickets. Parasitism rates in the North American fly species *Ormia ochracea*, which attacks several crickets of the genus *Gryllus* in the United States and a novel host on Hawaii, are low, generally not exceeding 16 percent (Adamo et al. 1995b, Cade 1975, Hedrick & Kortet 2006, Walker & Wineriter 1991, Zuk 1994). Female crickets are significantly less often parasitized than males, but they may still show infection levels of three to ten percent (Adamo et al. 1995b, Walker & Wineriter 1991, Zuk 1994). However, in Brazil 0.3 to 0.6 percent of female mole crickets (*Scapteriscus* 3 spp.) were found to be parasitized while no parasitized males were found (Fowler & Garcia 1987). There is an interesting difference between the cricket and the bushcricket systems. Female crickets may become parasitized if they walk through an area where a fly has recently deposited maggots (Adamo et al. 1995b), as can silent satellite males also (Cade 1975, Zuk 1994).

In the previous paragraph, I reviewed how seasonal occurrence differs between parasitoid-host systems. Even so large differences are observed in regard to parasitism rates between species in the field. What then are the reproductive patterns of Ormiini flies?

Brood sizes in field samples of five host-parasitoid systems are similar (Figure 14). Most hosts bear a single parasitoid larva and the distribution function is a negative exponential curve. This pattern is especially similar for the European *Therobia leonidei* attacking two bushcricket species and the American *Ormia ochracea* that parasitizes two cricket species. However, the brood size curve in the Australian *Homotrixa alleni* is rather flat and large brood sizes of seven and more parasitoids per host male are regularly found. This pattern is obviously the result of extended superparasitism by *Homotrixa alleni* (see discussion).

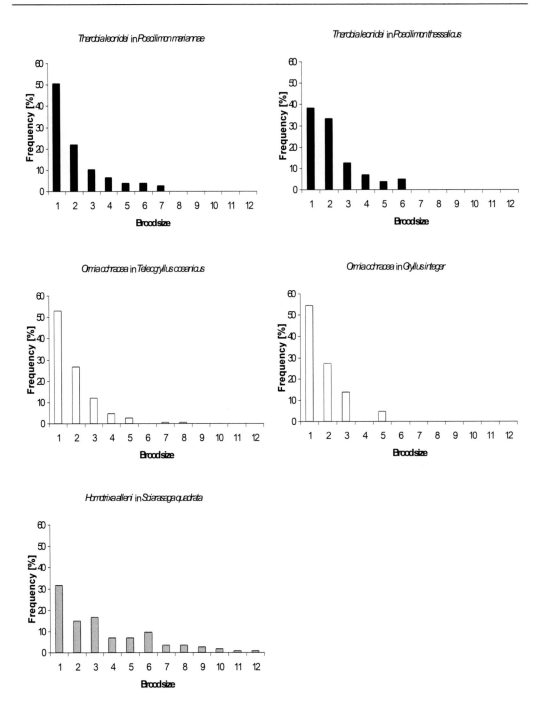

Figure 14. Frequency distribution of brood sizes of three different Ormiini parasitoids (*Therobia leonidei, Ormia ochracea* and *Homotrixa alleni)* in male bushcrickets in Europe (black, original data), Australia (grey, reanalyzed after Allen (1995a)) and in male crickets from North America (white, reanalyzed from Adamo et al. (1995b)) and from Hawaii, where the fly is introduced from North America and the non-original host introduced from Australia (white, reanalyzed from Kolluru & Zuk (2001)).

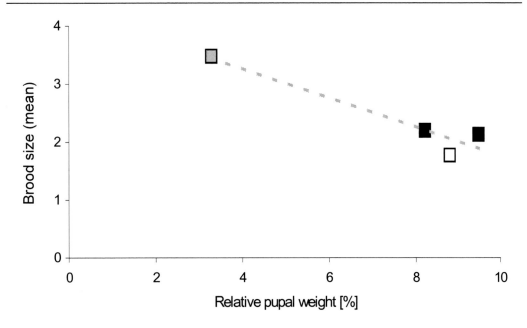

Figure 15. Correlation between the parasitoid mass (divided by host mass) and brood sizes in field samples. There is a strong negative correlation (y = -0.23x + 4.21, R^2 = 0.98), with relatively lighter parasitoids having more parasitoid maggots per host. Data of parasitoid flies of male bushcrickets in Europe (black, original data) and Australia (grey, reanalyzed after Allen (1995a)) and in male crickets from Hawaii (white, original data from Kolluru, pers. comm.).

Pupae of the European fly *Therobia leonidei* weigh on average 8.2 or 9.5 percent of their host males, depending on host species. This is similar to the relative mass of *Ormia ochracea* pupae, which weigh 8.8 percent of their host *Teleogryllus oceanicus* from Hawaii (Kolluru pers. comm.). *Ormia depleta* pupae obtained from artificially infected *Scapteriscus abbreviatus* weighed roughly five percent of host weight (Welch 2006). The Australian *Homotrixa* is once again an exception. Its pupae are twice as heavy as those from other Ormiini species. Nonetheless, *Homotrixa* hosts are so large that the relative parasitoid pupae mass 3.2 percent of host weight (Figure 15).

Classical clutch size theory (Lack 1947) predicts that a female should lay the number of eggs that maximizes her gain in fitness from the whole clutch. From field studies we can not be sure how many parasitoid females have contributed to the observed number of fly larvae within a host. However, the same fitness curve should apply regardless of whether a single or multiple females parasitize a host. Fitness curves for Ormiini flies are only available for *H. alleni* (Allen & Hunt 2001) but we can use the number of larvae to pupate per host as an approximation for other species (Figure 16). In all five systems the mean brood size in the field is much lower than the number of parasitoids that yields the maximum offspring number. This lower-than-optimal clutch size is a unifying pattern, independent of the response curves which are quite steep for *O. depleta*, *O. ochracea* and *H. alleni* or rather flat for *Therobia*.

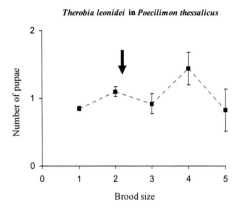

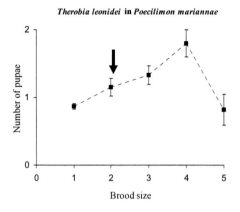

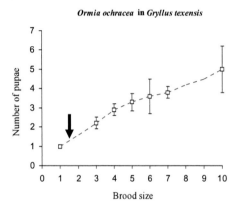

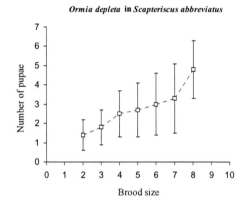

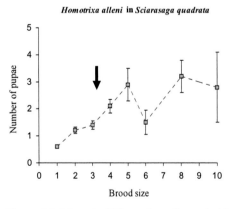

Figure 16. Mean (± SD) number of parasitoid pupae emerging from males of five different host species in relation to brood size. The arrow marks the mean brood size found in the field. The brood size data from the mole cricket *Scapteriscus abbreviatus* are from flying and not from grubbed individuals. This gave the possibility, that the real mean brood size is higher. Data of parasitoid flies in two bushcricket species in Europe (black, original data from chapter I) and one Australian bushcricket (grey, reanalyzed after Allen & Hunt (2001)). The data from *Ormia* fly species attacking crickets (white) and mole crickets were analyzed from Adamo et al. (1995b) and Welch (2006).

Oviposition Strategies

Tachinid flies show a diversity of oviposition strategies, even if they lack any specialized ovipositor. We can distinguish between species that lay eggs on the host (direct oviposition) versus those that lay eggs away from the host (indirect oviposition). The Ormiini show an advanced developing mode in that they exhibit ovovivipary, placing active first-instar planidia on or near to their hosts (Cade 1975, Adamo et al. 1995b, Allen et al. 1999). The indirect mode of oviposition is linked to a larger clutch size in *Ormia ochracea* (Adamo et al. 1995b), but the presence or absence of a host does not change the number of indirectly positioned planidia in *Homotrixa alleni* (Allen et al. 1999). We have no clear picture of how often or when female Ormiini use the direct or the indirect oviposition mode. However, fecundity seems to be quite similar between Ormiini flies (Allen & Hunt 2001, Fowler 1987b, Fowler & Garcia 1987, Kolluru & Zuk 2001, Wineriter & Walker 1990), with means of 148 to 219, ranging up to 600 planidia (see table 1). This number corresponds well with the summary given for tachinid flies having a direct-external oviposition mode using incubated eggs (Stireman et al. 2006).

Parasitoid fecundity also increases with body size - measured either as female head width (Allen & Hunt 2001) or female thorax length (Kolluru & Zuk 2001). The number of placed planidia may be three to four times higher than the brood size observed in cricket hosts (Adamo et al. 1995b, Kolluru & Zuk 2001). This does not necessarily mean there can be 75% larval mortality since some may infect by-passing females or satellite non-singing males (Adamo et al. 1995b, Zuk 1994).

DISCUSSION

Given the relatively small size of parasitoids, finding a suitable host might be a heavy task for most species. Hymenoptera are famous for their ability to use chemical cues either emitted directly from the host or from the host's microhabitat. Host plant odours are important as such chemicals (Godfray 1994). Ormiini flies in contrast rely on acoustic signals of their hosts for localization (Cade 1975) and act as illicit exploiters of the host communication system (Haynes & Yeargan 1999). Ormiini flies can very accurately hear (Mason et al. 2001) and attack hosts by homing in on their calling song (Müller & Robert 2001). Even while walking (Mason et al. 2005), *Ormia* flies are capable of distinguishing sound sources (Arthur & Hoy 2006). Signals derived directly from hosts are considered reliable indicators of their location and identity (Feener & Brown 1997). It is therefore not astonishing, that high levels of parasitism are found in most Ormiini-host systems studied (Allen 1995, Adamo et al. 1995b, Burk 1982, Kolluru & Zuk 2001), including my results on Greek bushcrickets presented here. Male parasitism in *P. mariannae* increased linear over the season and levelled between 50 and 57 percent in three consecutive years. This host species produces more complex and longer lasting (= poly-syllabic) songs compared to the monosyllabic *P. thessalicus*. This might explain largely the differences in the total amount of parasitized males (Lehmann & Heller 1998). However, substantial fluctuation exist in the parasitism rate between years in the monosyllabic *P. thessalicus*, similar to fluctuations between years found for another monosyllabic host *P. veluchianus* (Lehmann & Heller 1998). Because flies rely so strongly on sound as a host clue, a fluctuation might be related to differences in the amount the bushcricket males spend calling. Such variation in the time spend calling might be induced by environmental variables like weather conditions. Alternatively this levels are intrinsically linked to between years fluctuations in population densities of flies.

Once an adult fly has detected and localized a host, we expect a dynamic interaction between the host as the variable environment for the developing larvae and the decisions to be made by the mother. In general, oviposition behaviour in parasitoids is dynamic, responding to variation in host quality and availability (Papaj 2000). Parasitoids display remarkable inter- and intraspecific variation in their reproductive and associated traits. Adaptive explanations have been proposed for many of parasitism patterns and linked to key aspects of host ecology (Jervis et al. 2008). However, there are fundamental differences between the large body of results in Hymenoptera (e.g. Godfray 1994), and those in Diptera (Feener & Brown 1997), especially Tachinidae (Stireman et al. 2006).

Oviposition Strategies

Unlike parasitic Hymenoptera, tachinids lack a primitive piercing ovipositor. This missing prevents the injection of paralytic poisons, mutualistic polyDNA viruses, and other accessory substances. Instead of immobilizing their hosts and/or its immune system, tachinids allow them to continue to feed and grow while they develop inside (Stireman et al. 2006). Host performance decreases over time (Cade 1984, Adamo et al. 1995a, Allen 1995, Kolluru et al. 2002, Orozco & Bertram 2004, Lehmann & Lehmann 2006), probably in a response to the rapid growth of the maggots inside and the degrading effect on host physiology. Tachinidae flies can be distinguish according to their reproductive mode, especially whether they deposit eggs, larvae or even pupae (Meier et al. 1999). They can also be divided into species that oviposit directly on the host versus those that deposit in the surrounding of the host. These diverse oviposition strategies have evolved in concert with host-searching and attack strategies, changes in fecundity, and the types of hosts attacked (Stireman et al. 2006). Ormiini deposit active first-instar planidia (ovovivipary), flexible either direct on (Cade 1975) or in the surrounding of a host (Adamo et al. 1995b, Allen et al. 1999) and the planidia must gain entry into the host. The indirect mode of oviposition is connected to a larger clutch size in *Ormia ochracea* (Adamo et al. 1995b), but the presence or absence of a host does not change the number of indirectly positioned planidia in *Homotrixa alleni* (Allen et al. 1999). Field collected Ormiini females had up to 600 planidia, with means of 148 to 219 (Allen & Hunt 2001, Fowler 1987b, Fowler & Garcia 1987, Kolluru & Zuk 2001, Wineriter & Walker 1990) and larger wild-caught flies had higher fecundity (Allen & Hunt 2001, Kolluru & Zuk 2001).

Regarding the number of offspring deposited, Ormiini are gregarious parasitoids with mixed sex broods (Godfray 1994). Brood sizes are low compared to other tachinids, with a maximum of 12 breathing funnels observed in a single host (Allen 1995, Lehmann unpubl. data, Figure 14). The majority of hosts bear only a single parasitoid larvae. The brood size distribution is especially similar for the European *Therobia leonidei* attacking two bushcricket species and the American *Ormia ochracea* in two cricket species. Maggots in both Ormiini species reach similar maggots weights, with pupae from single infected hosts weighing in mean between 8.2 and 9.5 percent of host mass (Table 1). The large and heavy Australian bushcricket *Sciarasaga quadrata* however is parasitized by a larger number of maggots, but the relative parasitoid pupae mass is only 3.2 percent. This leads to the assumption of a correlation between relative parasitoid-to-host mass and the mean brood size (see Figure 15).

Ormiini may be unable to accurately access host size or quality, especially when females indirectly oviposit. The indirect infection mode further add to the infection uncertainty as some planidia might not reach the targeted singing male (Adamo et al. 1995b). The number of planidia placed around a targeted host is three to four times higher, than the brood size in field sampled cricket hosts (Adamo et al. 1995b, Allen et al. 1999). This means an approximated 75 percent of planidia mortality and therefore it is not astonishing, that no evidence of clutch sizes adjustment could be found, different to hymenopterans (Godfray 1994). Parasitized male crickets were slightly larger than unparasitized males (Kolluru & Zuk 2001), a result not found in a field experiment with bushcrickets (Lehmann et al. 2001). There was no correlation between host size and the number of larvae in all species tested (Table 1).

Larval Competition/Clutch Size Theory

One of the major decisions a gregarious parasitoid have to make is how many offspring to oviposit on an encountered host (Godfray 1994). Classical clutch size theory predicts, that a mother should lay the number of eggs, that maximizes her gain in fitness not from the single offspring, but from the whole clutch (Lack 1947). For insects that use discrete resources, such as parasitoids and seed feeders (see Amarillo-Suarez & Fox 2006), host size and host quality are major sources of phenotypic variation and constrain offspring growth (Hardy et al. 1992, Tsai et al. 2001). The fundamental assumption of parasitoid clutch size theory is that offspring fitness is a function of clutch size. There is a large body of evidence, that hymenopteran parasitoids adjust their clutch size to host size and host quality (Hardy et al. 1992, Godfray 1994, Mesterton-Gibbons et al. 2004, Traynor & Mayhew 2005, Häckermann et al. 2007).

Ormiini show a negative relationship between clutch size and progeny fitness, consistent with hymenopterans, but no adjustment of clutch size to host size (different from hymenopterans). The number of maggots placed on or around a singing host was similar between Ormiini species attacking larger (Allen et al. 1999) or smaller host species (Adamo et al. 1995b). Within species the number of maggots a host harboured was independent of host weight (Table 1), whereas between species the relatively heavier host species -the bushcricket *S. quadrata*- contained more parasitoids (Figure 15). However, this host survives longer after parasitoid attack (Allen 1995) than other Ormiini hosts (Lehmann & Lehmann 2006, Wineriter & Walker 1990), with an increased possibility, that the larger parasitoid number is the result of superparasitism (see discussion below) and not the outcome of clutch size adjustment by the parasitoid fly. Mixed results are obtained for pupal weight in regard to host weight. In my field studies, there was no relationship between parasitoid pupal weight and host weight (Figure 5), whereas in artificial infected mole crickets heavier hosts produced heavier fly maggots (Welch 2006). In phorid flies laboratory reared offspring were also positively related to the size of the host worker ant (Morrison & Gilbert 1998).

Larval competition in all studied Ormiini increased with brood size and led to reduced survival (Figure 9, Allen 1995b, Welch 2006) and pupal weight (Figure 7, Hunt & Allen 2000, Allen & Hunt 2001, Welch 2006) or pupal size (Adamo et al. 1995b, Kolluru & Zuk 2001). Hatching success of adult flies decreased with pupal weight in three Ormiini species

(Figure 8, Adamo et al. 1995b, Allen & Hunt 2001) and larger flies had higher fecundity (Allen & Hunt 2001, Kolluru & Zuk 2001). In conclusion, adult size and probably fitness is determined largely by the success in larval competition. The only available estimate of fitness in an Ormiini species was undertaken by Allen & Hunt (2001). To compare datasets, I used the number of larvae to pupate per host, instead of fitness curves. As in the hymenopterans (Godfray 1994), the observed brood size in the field was smaller than hosts can support to pupation in five Ormiini species (Figure 16). More precisely, between a third and half of all hosts contained only one parasitoid fly (Figure 14). There are many speculations why parasitoids may be selected to produce smaller clutches than the modelled optimum. One reason in Ormiini might be the fact that larger females survive better or are able to locate hosts better (Allen & Hunt 2001). A minimum size might be critical for hearing, given the fact, that the Ormiini ear is one of the smallest in the world (Lakes-Harlan & Heller 1992, Robert et al. 1992, 1998, Mason et al. 2001). A second explanation might be, that clutch size is hard to adjust to hosts, because the fly not always made contact with the targeted host. To have more planidia than optimum reaching and entering the host, will penalize all offspring with the risk of no survivors at all. This uncertainty over the number of planidia reaching a host undoubtly select for lower clutch sizes. It should also taken into account, that all measurements of offspring number and size are done using well-fed hosts in the laboratory. Environmental conditions might limit host performance and would influence the optimal parasitoid number. Given the broad host range, a moderate clutch size might also contribute to host flexibility and gave the flies the opportunity to invade new areas with alternative hosts of reduced size and weight. At least the cricket parasitoid *Ormia ochracea* shows a remarkable number of regional hosts over North America (Gray et al. 2007) and has even been able to adapt to a novel host after being introduced to Hawaii (Zuk 1994).

Sex Ratios

Males and females are produced in approximately equal numbers in most species with separate sex and any deviation from this equal sex ratio deserves explanation (Hardy 2002, Sheldon & West 2002). The better-studied hymenopteran parasitoids show sex-ratios varying with host size, host quality and local mate competition (Ode & Hunter 2002, Shuker et al. 2006). Most parasitic wasps are haplo-diploid, and ovipositing mothers can selectively fertilize eggs (Ode & Hunter 2002), whereas diplo-diploid species are constraint in there mechanisms of sex determination (Cook 2002, Cockburn et al. 2002, West & Sheldon 2002). In tachinids, the limited amount of studies did found equal sex ratios (Nakamura 1995), also in Ormiini (Adamo et al. 1995b, Allen & Hunt 2001). In phorid flies, females emerge from larger hosts (Morrison & Gilbert 1998) and laboratory studies revealed a pattern of sex ratio variation as a function of host size (Morrison et al. 1999). Sex ratios for single infected hosts were nearly equal for *Therobia leonidei* in my studies regardless of host species. Even if not statistically significant, I found a higher amount of female flies in larger brood sizes. This variation might not be the result of a deviated primary sex ratio controlled by the mothers, but due to differential mortality of the sexes in response to larval competition (e.g. Miura 2003). Larger brood sizes led to smaller offspring, so males in *Therobia leonidei* might be under a stronger selection for larger size than females. One possible explanation comes from the fact, that Ormiini form mating swarms (Lederhouse et al. 1976), like in a number of flies (Downes

1969). Therefore only large males are likely to gain access to females as in two Australian tachinids (Alcock & Smith 1995), leading to strong selection on male body size, levelling the ability to survive by small male maggots.

Superparasitism

Superparasitism is defined as the deposition of a clutch of eggs by a female parasitoid in or on a host, already parasitized by itself or a conspecific female (van Alphen & Visser 1990, Godfray 1994). Because the progeny of a superparasitizing female are normally at a competitive disadvantage relative to the progeny of the previous parasitoid, natural selection should favour females with the ability to discriminate against parasitized hosts. Such host discrimination have been widely documented for parasitic Hymenoptera, where parasitoids use a variety of cues to distinguish between parasitized and unparasitized hosts (Godfray 1994). Superparasitism in gregarious hymenopteran parasitoids can be an advantage adaptation, until the host carrying capacity is reached (Speirs et al. 1991, Dorn & Beckage 2007). In contrast to Hymenoptera, superparasitism appears to be a more regular phenomenon in parasitic Diptera (Feener & Brown 1997). This widespread occurrence of superparasitism in flies supports the impression, that most flies lack the ability to discriminate against parasitized hosts. Because female fitness is not only correlated to offspring quality but also to the total produced number of offspring, this lack of discrimination must not be a disadvantage. Parasitoid flies may optimizing their fitness by producing a lot of less fecund offspring (Nufio & Papaj 2004), most likely when host availability is low. Even so, the ability to discriminate between parasitized and unparasitized hosts may be less advantageous in species that attack agile or actively defended host, which is the case in carnivorous bushcrickets.

In the Ormiini, ovipositing females make not always direct contact with the host, and therefore they have little opportunity to recognize prior parasitization. Female Ormiini might discriminate against less favoured hosts by exploiting the information contained in hosts song. Indeed, Ormiini flies preferred the same song patterns as cricket host females, having higher chirp rates, longer chirp duration, and higher chirp amplitudes (Wagner 1996, Wagner & Basolo 2007). Host males more attractive for conspecific females also suffer from higher parasitism rates (Lehmann et al. 2001). Song production and structure (Cade 1984, Kolluru et al. 2002, Orozco & Bertram 2004, Lehmann & Lehmann 2006) and attractiveness of song (Lehmann & Lehmann 2006) decreases with the time since parasitism, so at least older parasitized hosts may be not superparasitized. The opportunity to use song for detecting superparasitism however, may be limited in the first days and Adamo et al. (1995b) found no evidence, that females of *O. ochracea* discriminate between unparasitized and parasitized cricket males. We have no data on direct observations of superparasitism in Ormiini, and it is ambiguous to determine, whether a field-collected host is superparasitized. However, several approaches are used to estimate superparasitism in gregarious species (Dorn & Beckage 2007). By classifying parasitoid maggots using their developmental stage inside the host, superparasitism was estimated to be between 4.7 (Kolluru & Zuk 2001) and 25 percent in north American field crickets (Adamo et al. 1995b). Using an threshold approach, I estimated an amount of 8.6 percent superparasitized males over the whole season, with a maximum of 17.6 percent in the bushcricket *P. thessalicus*. Superparasitism was obviously disadvantage, as never more than just one maggot hatched from this species. In conclusion, superparasitism might best be seen as nonadaptive in Ormiini, related to the non-contact oviposition and the lack of clear clues in the song to identify freshly parasitized hosts.

Multiparasitism

An other case of larval competition is found in cases where parasitized hosts are infected by a second species. Such multiparasitism was also reported between different species of Tachinidae (Iwao & Ohsaki 1996, Reitz 1996, Kan et al. 2003). Adult bushcrickets are found to be multiparasitized by tachinids and strepsipterans (Solulu et al. 1998), but both without the ability to hear their hosts. Even if Ormiini show broad host ranges, in no case were two Ormiini species found within one host. The only case of multiparasitism was observed between the Ormiini *H. alleni* and a sarcophagid fly species in the Australian bushcricket *S. quadrata* (Allen & Pape 1996). I also found Sarcophagidae as parasitoids of bushcrickets in Greece and Slovenia (Lehmann unpubl. data), but in those examples the sarcophagids must have entered the immature stages, as they hatched from nymphal field samples. In several hundred *Poecilimon* bushcrickets parasitized by the Ormiini *T. leonidei*, I never observed an other parasitoid species hatching. It is questionable why Sarcophagidae, in which at least one species have independently evolved a hearing organ (Lakes-Harlan et al. 1999, Robert et al. 1999) and parasitizes singing cicada in North America (Soper et al. 1976, Lakes-Harlan et al. 2000, Köhler & Lakes-Harlan 2001, 2003, Schniederkötter & Lakes-Harlan 2004, de Vries & Lakes-Harlan 2005) has not evolved the ability to exploit singing orthopterans as well. However, adult bushcrickets and crickets live quite hidden in their habitats, preventing probably most parasitoids to get access. The exploitation of acoustic signalling is therefore a private channel for Ormiini, which limits interspecific competition.

Host Ranges

Dipteran parasitoids generally attack a wider range of hosts compared to the more diverse parasitic wasps (Eggleton & Belshaw 1993). Many tachinid species are polyphagous, and a number of them show a striking variation in host ranges (Stireman et al. 2006). Host ranges in Ormiini are still poorly known (Lehmann 2003), so I must concentrate on the four systems studied in more detail. The comparative approach showed, that parasitism of bushcrickets is the ancestral stage and the switch to the ground living crickets and mole crickets is a derived pattern (Lehmann 2003), even if we lack a robust phylogeny for the flies. There is a strong difference between Ormiini parasitizing bushcrickets and those parasitizing crickets or mole crickets. The two bushcricket parasitoids *T. leonidei* and *H. alleni* have both broad host ranges (review in Lehmann 2003). Their hosts vary in regard to host feeding type (carnivore-herbivore), body size, and thus are from distantly related bushcricket subfamilies. The only limitation for parasitism seems to be the song pattern produced, but host species have either the male-only-calling or the female-responding communication system. It is therefore very likely, that much more host species will be found in the future, even in this well explored systems. This flexibility will also allow the flies to explore new or alternative host species easily, if only the song type fit into the physiological and anatomical properties of hearing.

However, little data exist for tropical areas, which would be a perfect test of our predictions. In general, males are the effected sex, with few exceptions, where acoustically responding females (Léonide 1969), non-singing females (Lehmann 2003) or even nymphae (Shapiro 1995), found as being parasitized.

Parasitism patterns differ in the two cricket-flies (including mole crickets). Their habit to shelter inside self-constructed burrows might protect crickets to some extend from the parasitoids. Therefore reported parasitism rates are lower than in bushcrickets (Adamo et al. 1995b, Cade 1975, Hedrick & Kortet 2006, Walker & Wineriter 1991, Zuk et al. 1993, Zuk 1994). Females are less often parasitized than males, but suffer to a much larger degree than bushcricket females (Adamo et al. 1995b, Walker & Wineriter 1991, Zuk 1994). Females and silent satellite male crickets may become parasitized, if they walk through an area, where a fly has recently deposited maggots (Adamo et al. 1995b, Cade 1975, Zuk 1994). The host ranges are more constrained, limited to closely related host species mostly from one genus. However, the fly *O. ochracea* uses crickets with locally differing song pattern (Gray et al. 2007) and has been adapted to a novel host after introduction into Hawaii (Zuk et al. 1993, Zuk 1994). In the cricket flies the hearing threshold is focussed on the song frequencies of crickets (Robert et al. 1992), restricting the parasitism to this group. Ormiini are obviously able to avoid an immune response during the first days, during which they remain in the thorax muscles (Léonide 1969, Adamo et al. 1995a). After the fly maggot have migrated into the abdomen, they turn the immune response of their host, to build the respiratory funnel from products of the host's immune response (Feener & Brown 1997). Hymenopteran parasitoids use a huge variety of mechanisms to escape the encapsulation process (Schmidt et al. 2001, Dubuffet et al. 2008), which obviously holds for Ormiini (Bailey & Zuk 2008) and allows them to infect a brood range of host species.

Phenotypic Plasticity

For insects that use discrete resources, such as parasitoids, host size is a major source of phenotypic variation and constrain offspring growth, influencing the evolution of body size and life history traits (Hardy et al. 1992, Tsai et al. 2001). Similar problems apply to seed feeders (Amarillo-Suarez & Fox 2006). Individuals in populations adapted to large hosts are generally larger than those adapted to small hosts, especially in species with scramble competition (Godfray 1994, Häckermann et al. 2007). In my study system, hatching maggots of the Ormiini *T. leonidei* in a population adapted to the larger host *P. mariannae* were much heavier than in a population adapted to the smaller host *P. thessalicus*. This difference in offspring weight is likely the response to large versus small hosts, in agreement with the usual expectations for scramble-competing species (Hardy et al. 1992, Tsai et al. 2001). Notably, maggots from the lighter host *P. thessalicus* were able to hatch at a much reduced minimal weight, in a range where no hatching was observed for maggots from the heavier host species. This plasticity likely buffers these flies from high mortality or low fitness, that they would otherwise experience, when encountering alternative host populations. Thus it provides the opportunity to infect a broad range of local host species, relatively independent of host body weight. Those adaptive plasticity should promote establishment and persistence in new environments (Ghalambor et al. 2007). The role of phenotypic plasticity is broadly covered by life history theory (Roff 2002), and such change in ecologically significant parameters can occur over tens of generations or fewer and is now widely documented in nature (Carroll et al. 2007). We need further experiments to disentangle the basis of such response, because much plasticity perceived as adaptation could be environmentally induced (nongenetic) plastic responses rather than (genetic) evolutionary adaptations (Gienapp et al. 2008). However, this is the first example I know of, that a fly parasitoid is found to respond in life history traits to its local host species.

Research Opportunities

The limitations for research within the Ormiini-host systems differ. Bushcrickets are easier to detect, at least in open habitats where the hosts can be sampled in reasonable numbers (Allen 1995, Burk 1982, Lakes-Harlan & Heller 1992, Lehmann & Heller 1998). The light colour of most bushcricket undersides is ideal for counting the melanized breathing funnels of the parasitoids (see Lehmann 2006). In contrast, crickets and mole cricket calls from inside of burrows are harder to find or capture (Walker & Wineriter 1991), and non-calling individuals might be underrepresented. Sound traps as time-effective alternatives for sampling face two serious problems. Firstly, parasitoid flies are also attracted along with their hosts, leading to parasitism inside traps (Walker & Wineriter 1991). Secondly, parasitism strongly affects host performance (Adamo et al. 1995a; Kolluru et al. 2002, 2004; Lehmann & Lehmann 2000a, 2000b, 2006), reducing calling duration and quality (Cade 1984, Kolluru et al. 2002, Lehmann & Lehmann 2006, Orozco & Bertram 2004). Even if it is not measured yet, host flight is an energy-demanding behaviour (Reinhold 1999) and so fewer parasitized hosts may be sampled. Therefore, parasitism rates and brood sizes might be underestimated in samples from sound traps. In the bushcricket systems, operating sound traps is technically more demanding, as the ultrasound song needs more sophisticated equipment for broadcasting. In my study species, *Therobia leonidei* no flies could be caught with different traps (similar to those described by Fowler 1988; Walker 1988, 1989) and flies could not be reared under semi-field conditions (Lehmann unpubl. data). Therefore any research on this fly must be field-based and limits the experimental possibilities.

To date, Ormiini research has yielded interesting results, regarding host selection, host switching from bushcrickets to crickets and the observation, that the uncertainty of host parasitism limits the potential for clutch size adaptation in this group. Still far more results are available for hymenopteran parasitoids, but flies and especially tachinids might be a good addition to understand general trends in parasitoid-host relationships. Ormiini form only a small group, but due to their influential position in evolutionary biology (Zuk & Kolluru 1998) and biological control (Parkman et al. 1996, Walker et al. 1996), we know more about this small group than about most other parasitoid flies (Feener & Brown 1997, Stireman et al. 2006). The potential for further research is therefore manifold and I encourage future researchers to study this group, especially results from tropical areas have the potential to contribute substantially to the understanding of host-parasitoid relationships and that of illicit receivers of acoustic communication in particular.

ACKNOWLEDGEMENT

I thank Robert Hickson and Arne Lehmann for discussion of the manuscript. I also thank Howard Frank and Gita Kolluru for providing unpublished data and answering questions. My own research on Ormiini was supported by the Deutsche Forschungsgemeinschaft (DFG), the Deutscher Akademischer Austauschdienst (DAAD), Ethologische Gesellschaft and Konrad-Adenauer-Stiftung (KAS).

REFERENCES

Adamo SA, Robert D, Hoy RR (1995a): Effects of a tachinid parasitoid, *Ormia ochracea*, on the behaviour and reproduction of its male and female field cricket hosts (*Gryllus* spp.). *Journal of Insect Physiology* 41: 269-277.

Adamo SA, Robert D, Perez J, Hoy RR (1995b): The response of an insect parasitoid, *Ormia ochracea* (Tachinidae), to the uncertainty of larval success during infestation. *Behavioral Ecology and Sociobiology* 36: 111-118.

Alcock J, Smith AP (1995): Landmark-defence and scramble competition mating systems in two Australian tachinid flies (Diptera). *Journal of the Kansas Entomological Society* 68:85–94.

Allen GR (1995): The biology of the phonotactic parasitoid, *Homotrixa* sp. (Diptera: Tachinidae), and its impact on the survival of male *Sciarasaga quadrata* (Orthoptera: Tettigoniidae) in the field. *Ecological Entomology* 20: 103-110.

Allen GR (1998): Diel calling activity and field survival of the bushcricket, *Sciarasaga quadrata* (Orthoptera: Tettigoniidae): a role for sound-locating parasitic flies? *Ethology* 104: 645-660.

Allen GR, Hunt CJ (2001) Larval competition, adult fitness, and reproductive strategies in the acoustically orienting Ormiine *Homotrixia alleni* (Diptera: Tachinidae). *Journal of Insect Behavior* 14:283-296

Allen GR, Pape T (1996): Description of female and biology of *Blaesoxipha ragg* Pape (Diptera: Sarcophagidae), a parasitoid of *Sciarasaga quadrata* Rentz (Orthoptera: Tettigoniidae) in Western Australia. *Australian Journal of Entomology* 35: 147-151.

Allen GR, Kamien D, Berry O, Byrne P, Hunt J (1999): Larviposition, host cues, and planidial behavior in the sound-locating parasitoid fly *Homotrixa alleni* (Diptera: Tachinidae). *Journal of Insect Behavior* 12: 67-79.

Amarillo-Suárez AR, Fox CW (2006): Population differences in host use by a seed-beetle: local adaptation, phenotypic plasticity and maternal effects. *Oecologia* 150: 247-258.

Arthur BJ, Hoy RR. (2006): The ability of the parasitoid fly *Ormia ochracea* to distinguish sounds in the vertical plane. *Journal of the Acoustical Society of America* 120: 1546-1549.

Bailey NW, Zuk M (2008): Changes in immune effort of male field crickets infested with mobile parasitoid larvae. *Journal of Insect Physiology* 54: 96-104.

Bell HA, Marris GC, Prickett AJ, Edwards JP (2005): Influence of host size on the clutch size and developmental success of the gregarious ectoparasitoid *Eulophus pennicornis* (Nees)

(Hymenoptera: Braconidae) attacking larvae of the tomato moth *Lacanobia oleracea* (L.) (Lepidoptera: Noctuidae). *Journal of Experimental Biology* 208: 3199-3209.

Burk T (1982): Evolutionary significance of predation on sexually signalling males. *Florida Entomologist* 65: 90-104.

Cade WH (1975): Acoustically orienting parasitoids: fly phonotaxis to cricket song. *Science* 190: 1312-1313.

Cade WH (1981): Alternative male strategies: genetic differences in crickets. *Science* 212: 563-564.

Cade WH (1984): Effects of fly parasitoids on nightly calling duration in field crickets. *Canadian Journal of Zoology* 62: 226-228.

Cade WH, Ciceran M, Murray A-M (1996): Temporal patterns of parasitoid fly (*Ormia ochracea*) attraction to field cricket song (*Gryllus integer*). *Canadian Journal of Zoology* 74: 393-395.

Carroll SP, Hendry AP, Reznick DN, Fox CW (2007): Evolution on ecological time-scales. *Functional Ecology* 21: 387-393.

Cockburn A, Legge S, Double MC (2002): Sex ratios in birds and mammals: can the hypotheses be disentangled? In: Hardy ICW (ed): *Sex ratios - concepts and research methods*. Cambridge University Press, Cambridge, pp 266-286.

Cook JM (2002): Sex determination in invertebrates. In: Hardy ICW (ed): *Sex ratios - concepts and research methods*. Cambridge University Press, Cambridge, pp 178-194.

de Vries T, Lakes-Harlan R (2005): Phonotaxis of *Emblemasoma auditrix* (Sarcophagidae, Diptera) in relation to number of larvae and age. *Zoology* 108: 239-246.

de Vries T, Lakes-Harlan R (2007): Prenatal cannibalism in an insect. *Naturwissenschaften* 94: 477-482.

Dorn S, Beckage NE (2007): Superparasitism in gregarious hymenopteran parasitoids: ecological, behavioural and physiological perspectives. *Physiological Entomology* 32: 199-211.

Downes JA (1969): The swarming and mating flight of Diptera. *Annual Review of Entomology* 14: 271-298.

Dubuffet A, Doury G, Labrousse C, Drezen J-M, Carton Y, Poirié M (2008): Variation of success of *Leptopilina boulardi* in *Drosophila yakuba*: *The mechanisms explored*. *Developmental and Comparative Immunology* 32: 597-602.Eggleton P, Belshaw R (1992): *Insect parasitoids: an evolutionary overview*. Philosophical Transactions of the Royal Society of London, Series B 337: 1-20.

Eggleton P, Belshaw R (1993): Comparisons of dipteran, hymenopteran and coleopteran parasitoids: provisional phylogenetic explanations. *Biological Journal of the Linnean Society* 48: 213-226.

Feener DH Jr., Brown BV (1997): Diptera as parasitoids. *Annual Review of Entomology* 42: 73-97.

Fowler HG (1987a): Field confirmation of the phonotaxis of *Euphasiopteryx depleta* (Diptera: Tachinidae) to calling males of *Scapteriscus vicinus* (Orthoptera: Gryllotalpidae). *Florida Entomologist* 70: 409-410.

Fowler HG (1987b): Field behaviour of *Euphasiopteryx depleta* (Diptera: Tachinidae): phonotactically orienting parasitoids of mole crickets (Orthoptera: Gryllotalpidae: *Scapteriscus*). *Journal of the New York Entomological Society* 95: 474-480.

Fowler HG (1988): Traps for collecting live *Euphasiopteryx depleta* (Diptera: Tachinidae) at a sound source. *Florida Entomologist* 71: 654-656.

Fowler HG, Garcia CR (1987): Attraction to synthesized songs and experimental and natural parasitism of *Scapteriscus* mole crickets (Orthoptera: Gryllotalpidae) by *Euphasiopteryx depleta* (Diptera: Tachinidae). *Rev. Brasil. Biol.* 47: 371-374.

Fox CW, Czesak ME (2000): Evolutionary ecology of progeny size in arthropods. *Annual Reviews in Entomology* 45: 341-369.

Ghalambor CK, McKay JK, Carroll SP, Reznick DN (2007): Adaptive versus non-adaptive phenotypic plasticity and the potential for contemporary adaptation in new environments. *Functional Ecology* 21: 394-407.

Gienapp P, Teplitsky C, Alho JS, Mills JA, Merilä J (2008): Climate change and evolution: disentangling environmental and genetic responses. *Molecular Ecology* 17: 167-178.

Godfray HCJ (1994): *Parasitoids: behavioral and evolutionary ecology*. Princeton University Press. Princeton, New Jersey.

Godfray HCJ, Patridge L, Harvey PH (1991): *Clutch size. Annual Review of Ecology* 22: 409-429.

Gray DA, Banuelos CM, Walker SE, Cade WH, Zuk M (2007): Behavioural specialization among populations of the acoustically orienting parasitoid fly *Ormia ochracea* utilizing different cricket species as hosts. *Animal Behaviour* 73: 99-104.

Häckermann J, Rott AS, Dorn S (2007): How two different host species influence the performance of a gregarious parasitoid: host size is not equal to host quality. *Journal of Animal Ecology* 76: 376-383.

Hardy ICW, Griffiths NT, Godfray HCJ (1992): Clutch size in a parasitoid wasp: a manipulation experiment. *Journal of Animal Ecology* 61: 121-129.

Hardy ICW (2002): *Sex ratios - concepts and research methods*. Cambridge University Press, Cambridge.

Hassell MP (2000): *The spatial and temporal dynamics of host-parasitoid interactions*. Oxford University Press. Oxford, Massachusetts.

Haynes KF, Yeargan KV (1999): Exploitation of intraspecific communication systems: illicit signalers and receivers. *Annals of the Entomological Society of America* 92: 960-970.

Hedrick AV, Kortet R (2006): Hiding behaviour in two cricket populations that differ in predation pressure. *Animal Behaviour,* 72: 1111-1118.

Hochberg ME, Ives AR (2000): *Parasitoid population biology*. Princeton University Press. Princeton, New Jersey.

Hunt J, Allen GR (2000):. Larval density and developmental instability in the acoustically orienting parasitoid *Homotrixa alleni*. *Acta Ethologica* 2: 129-132.

Iwao K, Ohsaki N (1996): Inter- and intraspecific interactions among larvae of specialist and generalist parasitoids. *Researches on Population Ecology* 38: 265-273.

Jervis MA, Ellers J, Harvey JA (2008): Resource acquisition, allocation, and utilization in parasitoid reproductive strategies. *Annual Review of Entomology* 53: 361-385.

Kan E, Fukuhara N, Hidaka T (2003): Parasitism by tachinid parasitoids (Diptera: Tachinidae) in connection with their survival strategy. *Applied Entomology and Zoology* 38: 131-140.

Köhler U, Lakes-Harlan R (2001): Auditory behaviour of a parasitoid fly (*Emblemasoma auditrix*, Diptera, Sarcophagidae). *Journal of Comparative Physiology* A 187: 581-587.

Köhler U, Lakes-Harlan R (2003): Influence of habitat structure on phonotactic strategy of a parasitoid fly (*Emblemasoma auditrix*, Diptera, Sarcophagidae). *Ecological Entomology* 28: 758-765.

Kolluru GR (1999): Variation and repeatability of calling behavior in crickets subject to a phonotactic parasitoid fly. *Journal of Insect Behavior* 12: 611–627.

Kolluru GR, Zuk M (2001): Parasitism patterns and the size-fecundity relationship in the acoustically-orienting dipteran parasitoid *Ormia ochracea*. *Canadian Journal of Zoology* 79: 973-979.

Kolluru GR, Zuk M, Chappell MA (2002): Reduced reproductive effort in male field crickets infested with parasitoid fly larvae. *Behavioral Ecology* 13:607-614.

Kolluru GR, Chappell MA, Zuk M (2004): Sex differences in insect metabolic rates: field crickets (*Teleogryllus oceanicus*) and their dipteran parasitoids (*Ormia ochracea*). *Journal of Comparative Physiology* B 174: 641-648.

Lack D (1947): *The significance of clutch size,* I–II. Ibis 89: 302-352.

Lakes-Harlan R, Heller K-G (1992): Ultrasound-sensitive ears in a parasitoid fly. *Naturwissenschaften* 79: 224-226.

Lakes-Harlan R, Stölting H, Moore TE (2000): Phonotactic behaviour of a parasitoid fly (*Emblemasoma auditrix*, Diptera, Sarcophagidae) in response to the calling song of its host cicada (*Okanagana rimosa*, Homoptera, Cicadidae). *Zoology* 103: 31-39.

Lakes-Harlan R, Stölting H, Stumpner A (1999): Convergent evolution of insect hearing organs from a preadaptive structure. *Proceedings of the Royal Society of London* B 266: 1161-1167.

Lederhouse RC, Morse RA, Ambrose JT, Burgett DM, Conner WE, Edwards L, Fell RD, Rutowski R, Turell M (1976): Crepuscular mating aggregations in certain *Ormia* and *Sitophaga*. *Annals of the Entomological Society of America* 69: 656-658.

Lehmann AW (1998): *Artbildung, akustische Kommunikation und sexuelle Selektion bei griechischen Laubheuschrecken der Poecilimon propinquus-Gruppe (Orthoptera: Phaneropteridae).* PhD University Erlangen-Nürnberg.

Lehmann AW, Willemse F, Heller K-G (2006): *Poecilimon gerlindae* nov. spec. – a new bushcricket of the *Poecilimon propinquus*-group from Greece (Orthoptera: Phaneropteridae). Articulata 21: 109-119.

Lehmann G (1998): *Der Einfluß der phonotaktischen, parasitoiden Fliege Therobia leonidei (Tachinidae: Ormiini) auf die akustische Kommunikation von Laubheuschrecken (Tettigonioidea, Phaneropteridae).* PhD University Erlangen-Nürnberg. 111 pp.

Lehmann GUC (2003): Review of biogeography, host range and evolution of acoustic hunting in Ormiini (Insecta, Diptera, Tachinidae), parasitoids of night-calling bushcrickets and crickets (Insecta, Orthoptera, Ensifera). *Zoologischer Anzeiger* 242: 107-120.

Lehmann GUC (2006): The life cycle of *Therobia leonidei* Mesnil, 1964 – an European parasitoid of bushcrickets (Tachinidae, Ormiini). *Tachinid Times* 19: 8-10.

Lehmann GUC, Heller K-G (1998): Bushcricket song structure and predation by the acoustically orienting parasitoid fly *Therobia leonidei* (Diptera: Tachinidae: Ormiini). *Behavioral Ecology and Sociobiology* 43: 239-245.

Lehmann GUC, Lehmann AW (2000a): Spermatophore characteristics in bushcrickets vary with parasitism and remating interval. *Behavioral Ecology and Sociobiology* 47: 393-399.

Lehmann GUC, Lehmann AW (2000b): Female bushcrickets mated with parasitized males show rapid remating and reduced fecundity (Orthoptera: Phaneropteridae: *Poecilimon mariannae*). *Naturwissenschaften* 87: 404-407.

Lehmann GUC, Lehmann AW (2006): Potential lifetime reproductive success for male bushcrickets following parasitism by a phonotactic fly. *Animal Behaviour* 71: 1103-1110.

Lehmann GUC, Lehmann AW (in press): Variation in body size between populations of the bushcricket *Poecilimon thessalicus* Brunner von Wattenwyl, 1871 (Orthoptera: Phaneropteridae): an ecological adaptation? *Journal of Orthoptera Research*

Lehmann GUC, Heller K-G, Lehmann AW (2001): Male bushcrickets favoured by parasitoid flies when acoustically more attractive for conspecific females (Orthoptera: Phaneropteridae / Diptera: Tachinidae). *Entomologia Generalis* 25: 135-140.

Léonide JC (1969): Recherches sur la biologie de divers diptères endoparasites d'Orthoptères. Memoires du Museum National d'histoire Naturelle serie A, *Zoologie* 53: 1–246.

Mackauer M, Chau A (2001): Adaptive self superparasitism in a solitary parasitoid wasp: the influence of clutch size on offspring size. *Functional Ecology* 15: 335-343.

Mangold JR (1978): Attraction of *Euphasiopteryx ochracea*, *Corethrella* sp. and gryllids to broadcast songs of the southern mole cricket. Florida. *Entomologist* 61: 57-61.

Mason AC, Oshinsky ML, Hoy RR (2001): Hyperacute directional hearing in a microscale auditory system. *Nature* 410: 686-690.

Mason AC, Lee N, Oshinsky ML (2005): The start of phonotactic walking in the fly *Ormia ochracea*: a kinematic study. *Journal of Experimental Biology* 4: 699-708.

Mayhew PJ (1998): Offspring size-number strategy in the bethylid parasitoid *Laelius pedatus*. *Behavioral Ecology* 9: 54-59.

Meier R, Kotbra M, Ferrar P (1999): Ovoviviparity and viviparity in the Diptera. *Biological Reviews* 74: 199-258.

Mesterton-Gibbons M, Hardy ICW (2004): The influence of contests on optimal clutch size: a game-theoretic model. *Proceedings of the Royal Society of London B* 271: 971-978.

Miura, K (2003): Parasitism of *Parapodisma* grasshopper species by the flesh fly, *Blaesoxipha japonensis* (Hori) (Diptera : Sarcophagidae). *Applied Entomology and Zoology* 38: 537-542.

Morrison LW, Gilbert LE (1998): Parasitoid-host relationships when host size varies: the case of *Pseudacteon* flies and *Solenopsis* fire ants. *Ecological Entomology* 23: 409-416.

Morrison LW, Porter SD, Gilbert LE (1999): Sex ratio variation as a function of host size in *Pseudacteon* flies (Diptera, Phoridae), parasitoids of *Solenopsis* fire ants (Hymenoptera, Formicidae). *Biological Journal of the Linnean Society* 66: 257-267.

Müller P, Robert D (2001): A shot in the dark: the silent quest of a free-flying phonotactic fly. *Journal of Experimental Biology* 204: 1039-1052.

Nakamura S (1995): Optimal clutch size for maximizing reproductive success in a parasitoid fly, *Exorista japonica* (Diptera: Tachinidae). *Applied Entomology and Zoology* 30: 425-431.

Nufio CR, Papaj DR. (2004) Superparasitism of larval hosts by the walnut fly, *Rhagoletis juglandis*, and its implications for female and offspring performance. *Oecologia* 141: 460-467.

Ode PJ, Hunter MS (2002): Sex ratios of parasitic Hymenoptera with unusual life-histories. In: Hardy ICW (ed): *Sex ratios - concepts and research methods*. Cambridge University Press, Cambridge, pp 218-234.

Orozco SX, Bertram SM (2004): Parasitized male field crickets exhibit reduced trilling bout rates and durations. Ethology 110: 909-917.

Papaj DR (2000): Ovarian dynamics and host use. *Annual Review of Entomology* 45: 423-448.

Parkman JP, Frank JH, Walker TJ, Schuster DJ (1996): Classical biological control of *Scapteriscus* spp. (Orthoptera: Gryllotalpidae) in Florida. *Environmental Entomology* 25: 1415-1420.

Reinhold K. (1999): Energetically costly behaviour and the evolution of resting metabolic rate in insects. *Functional Ecology* 13: 217-224.

Reitz AR (1996): Interspecific competition between two parasitoids of *Helicoverpa zea: Eucelatoria bryani* and *E. rubentis*. *Entomologia Experimentis et Applicata* 79: 227-234.

Reitz SR, Adler PH (1995): Fecundity and oviposition of *Eucelatoria bryani*, a gregarious parasitoid of *Helicoverpa zea* and *Heliothis virescens*. *Entomologia Experimentis et Applicata* 75: 175-181.

Robert D, Amoroso J, Hoy RR (1992): The evolutionary convergence of hearing in a parasitoid fly and its cricket host. *Science* 258: 1135-1137.

Robert D, Miles RN, Hoy RR (1998): Tympanal mechanics in the parasitoid fly *Ormia ochracea*: intertympanal coupling during mechanical vibration. *Journal of Comparative Physiology* A 183: 443-452.

Robert D, Miles RN, Hoy RR (1999): Tympanal hearing in the sarcophagid parasitoid fly *Emblemasoma* sp.: the biomechanics of directional hearing. *Journal of Experimental Biology* 202: 1865-1876.

Roff DA (2002): *Life history evolution*. Sinauer Associates, Sunderland.

Rotenberry, J. T., Zuk, M., Simmons, L. W. & Hayes, C. 1996 Phonotactic parasitoids and cricket song structure: an evaluation of alternative hypotheses. *Evol. Ecol.* 10, 233–243.

Schmidt O, Theopold U, Strand M (2001): Innate immunity and its evasion and suppression by hymenopteran endoparasitoids. *BioEssays* 23: 344-351.

Schniederkötter K, Lakes-Harlan R (2004): Infection behaviour of a parasitoid fly (*Emblemasoma auditrix*, Diptera, Sarcophagidae). *Journal of Insect Science* 4: 36 [7 pp.]

Shapiro L (1995): Parasitism of *Orchelimum* katydids (Orthoptera: Tettigoniidae) by *Ormia lineifrons* (Diptera: Tachinidae). *Florida Entomologist* 78: 615-616.

Sheldon BC, West SA (2002): Sex ratios. In: Pagel M (ed) *Encyclopedia of Evolution,* Oxford University Press, Oxford, pp 1040-1044.

Shuker DM, Pen I, West SA (2006): Sex ratios under asymmetrical local mate competition in the parasitoid wasp *Nasonia vitripennis*. *Behavioral Ecology* 17: 345–352.

Solulu TM, Simpson SJ, Kathirithamby (1998): The effect of strepsipteran parasitism on a tettigoniid pest of oil palm in Papua New Guinea. *Physiological Entomology* 23: 388-398.

Soper RS, Shewell GE, Tyrell D (1976): *Colcondamyia auditrix* nov. sp. (Diptera: Sarcophagidae), a parasite which is attracted by the mating song of its host, *Okanagana rimosa* (Homoptera: Cicadidae). *Canadian Entomologist* 108: 61-68.

Speirs DC, Sherratt TN, Hubbard SH (1991): Parasitism diets: does superparasitism pay? *Trends in Ecology and Evolution* 6: 22-25.

Stireman JO III, O'Hara JE, Wood DM (2006): Tachinidae: evolution, behavior, and ecology. *Annual Review Entomology* 2006. 51:525–55

Stumpner A, Allen GR, Lakes-Harlan R (2007): Hearing and frequency dependence of auditory interneurons in the parasitoid fly *Homotrixa alleni* (Tachinidae: Ormiini). *Journal of Comparative Physiology* A 193: 113-125.

Traynor RE, Mayhew PJ (2005): A comparative study of body size and clutch size across the parasitoid Hymenoptera. *Oikos* 109 (2), 305–316.

Tsai ML, Li JJ, Dai CF (2001): How host size may constrain the evolution of parasite body size and clutch size. The parasitic isopod *Ichthyonexus fushanensis* and its host fish, *Varicorhinus bacbatulus*, as an example. *Oikos* 92: 13–19.

van Alphen JJM, Visser ME (1990): Superparasitism as an adaptive strategy for insect parasitoids. *Annual Review of Entomology* 35: 59-79.

Wagner WE (1996): Convergent song preferences between female field crickets and acoustically orienting parasitoid flies. *Behavioral Ecology* 7: 279–285.

Wagner WE Jr., Basolo A (2007): Host preferences in a phonotactic parasitoid of field crickets: the relative importance of host song characters. *Ecological Entomology* 32: 478–484.

Wajnberg E, Bernstein C, van Alphen J (2007): *Behavioral ecology of insect parasitoids: From theoretical approaches to field applications.* Blackwell Publishing, 464 pp.

Walker TJ (1986): Monitoring the flights of field crickets (*Gryllus* spp.) and a tachinid fly (*Euphasiopteryx ochracea*) in north Florida. *Florida Entomologist* 69: 678-685.

Walker TJ (1988): Acoustic traps for agriculturally important insects. *Florida Entomologist* 71: 484-492.

Walker TJ (1989): A live trap for monitoring *Euphasiopteryx* and tests with *E. ochracea* (Diptera: Tachinidae). *Florida Entomologist* 72: 314-319.

Walker TJ, Wineriter SA (1991): Hosts of a phonotactic parasitoid and levels of parasitism (Diptera: Tachinidae: *Ormia ochracea*). *Florida Entomologist* 74: 554-559.

Walker TJ, Parkman JP, Frank JH, Schuster DJ (1996): Seasonality of *Ormia depleta* and limits to its spread. *Biological Control* 6: 378-383.

Welch CH (2006): Intraspecific competition for resources by *Ormia depleta* (Diptera: Tachinidae) larvae. *Florida Entomologist* 89: 497-501.

West SA, Sheldon BC (2002): Constraints in the evolution of sex ratio adjustment. *Science* 295: 1685-1688.

Willemse F, Heller K-G (1992): Notes on systematics of greek species of *Poecilimon* Fischer,1853 (Orthoptera: Phaneropteridae). *Tijdschrift voor Entomologie* 135: 299-315.

Wineriter SA, Walker TJ (1990): Rearing phonotactic parasitoid flies (Diptera: Tachinidae, Ormiini, *Ormia* spp.). *Entomophaga* 35: 621-632.

Zaviezo T, Mills NJ (2000): Factors influencing the evolution of clutch size in a gregarious insect parasitoid. *Journal of Animal Ecology* 69: 1047-1057.

Zuk M (1994): Singing under pressure: phonotactic parasitoid flies in Hawaiian cricket hosts. *Field Notes* 10:477-484.

Zuk M, Kolluru GR (1998): Exploitation of sexual signals by predators and parasitoids. *Quarterly Review Biology* 73: 415-438.

Zuk M, Simmons LW, Cupp L (1993): Calling characteristics of parasitized and unparasitized populations of the field cricket *Teleogryllus oceanicus*. *Behavioral Ecology and Sociobiology* 33: 339-343.

Zuk M, Simmons LW, Rotenberry JT (1995): Acoustically-orienting parasitoids in calling and silent males of the field cricket *Teleogryllus oceanicus*. *Ecological Entomology* 20: 380-383.

Zuk M, Rotenberry JT, Simmons LW (1998): Calling songs of field crickets (*Teleogryllus oceanicus*) with and without phonotactic parasitoid infection. *Evolution* 52: 166-171.

Zuk M, Rotenberry JT, Simmons LW (2001): Geographical variation in calling song of the field cricket *Teleogryllus oceanicus*: the importance of spatial scale. *Journal of Evolutionary Biology* 14: 731-741.

Zuk M, Rotenberry JT, Tinghitella RM (2006): Silent night: adaptive disappearance of a sexual signal in a parasitized population of field crickets. *Biology Letters* 2: 521-524.

In: Animal Behavior: New Research
Editors: E. A. Weber, L. H. Krause

ISBN: 978-1-60456-782-3
© 2008 Nova Science Publishers, Inc.

Chapter 6

INSIGHTS INTO THE ACOUSTIC BEHAVIOUR OF POLAR PINNNIPEDS – CURRENT KNOWLEDGE AND EMERGING TECHNIQUES OF STUDY

Ilse Van Opzeeland, Lars Kindermann, Olaf Boebel and Sofie Van Parijs
Alfred Wegener Institute, Bremerhaven, Germany

ABSTRACT

This chapter will provide a review of the acoustic behaviour of polar pinnipeds. It will also present a detailed update of new and emerging passive acoustic technologies and how these can further the study of behaviour for polar marine mammals.

Both Arctic and Antarctic pinnipeds are known to exhibit a range of adaptations which enable them to survive and reproduce in an ice-dominated environment. However, large gaps still exist in our understanding of the fundamental ecology of these species, as investigations are severely hampered by the animals' inaccessibility. Improving our understanding of ice-breeding species and the effects that changes in habitat might have on their behaviour is vital, as current climatic trends are rapidly altering the polar environments.

For pinnipeds, acoustic communication is known to play an important role in various aspects of their behaviour. Mother-pup reunions and the establishment of underwater territories during the mating season are examples which, for the majority of species, are known to be mediated by vocal signalling. Acoustic measurements therefore provide an essential tool to study ice-breeding pinnipeds as recordings can be used to remotely monitor sounds, track animal movements and determine seasonal changes in movements and distribution. Recent advances in recording technologies, now allow the acquisition of continuous long-term acoustic data sets, even from the remotest of polar regions.

To date, a range of different types of passive acoustic instruments are used, the choice of which depends largely on the purpose of the study. These instruments in addition to computer-based methods that have been developed for automated detection, classification and localization of marine mammal sounds will be discussed. The autonomous PerenniAL Acoustic Observatory in the Antarctic Ocean (PALAOA) is presented here as one example of such recording systems.

INTRODUCTION

Dispersal of pinnipeds into polar areas is thought to have begun with the evolution of large body size in the ancestral pinnipeds (Costa 1993). In Costa's (1993) model, early pinnipeds exhibited a primitive form of otariid breeding patterns with females requiring numerous short duration foraging trips to sustain lactation. The evolution of large body size enabled females to separate foraging from lactation as females had increased maternal reserves to rely on. This separation of foraging and breeding is thought to finally have enabled these basal phocids to inhabit and reproduce in less-productive areas such as the Atlantic, in relative absence of resource competitors. Upon reaching higher latitudes, the shortened lactation period would have pre-adapted these early phocids to breeding on unstable substrates, such as ice (Costa 1993). The establishment of ice-breeding along with development of a mainly aquatic life style has influenced many aspects of pinniped behaviour in high latitude habitats e.g. the timing of reproduction, duration of lactation period and the development of different mating strategies. Although ice-breeding pinnipeds share many similarities in behavioural patterns, there is also considerable variance in the social and physical conditions of their breeding habitat (e.g. Lydersen & Kovacs 1999). To date however, clear gaps still exist in what is known about the behaviour of polar pinnipeds. This is due to the fact that behavioural studies in polar pinniped species are severely hampered since many of the behavioural patterns are likely to take place in offshore waters or in remote pack-ice areas, logistically limiting the possibilities of study. All of the ice-breeding pinniped species are however, known to produce underwater vocalizations, the monitoring of which has proved to be a valuable tool to study many aspects of pinniped behaviour (e.g. Thomas & DeMaster 1982; Rogers et al. 1996; Van Parijs et al. 1997; Perry & Terhune 1999; Van Parijs et al. 2003; Van Parijs et al. 2004; Terhune & Dell'Apa 2006; Rouget et al. 2007). This review aims to provide an overview of the existing information and current gaps in knowledge on the acoustic behaviour of ice-breeding pinnipeds. Additionally, we summarize recent developments in recording technologies, which now enable acquisition of long-term acoustic data sets in remote areas and evaluate how these techniques may contribute to the improvement of our fundamental understanding of the behaviour of polar pinnipeds.

POLAR HABITATS

The Southern Ocean surrounds the Antarctic continent. The northern boundary of the Southern Ocean is formed by the Antarctic Convergence or the southern polar frontal zone, which forms a sharp temperature boundary between northern temperate waters and southern polar waters. The polar front is an important factor in the distribution of marine mammals, as it defines the southern extent of tropical and temperate species. There are four species of ice-breeding seals in the Antarctic, all of which occupy different niches in the sea ice habitat:

leopard (*Leptonyx hydrurga*), crabeater (*Lobodon carcinophagus*), Ross (*Ommatophoca rossii*) and Weddell seal (*Leptonychotes weddellii*). The Antarctic sea-ice habitat differs substantially from that in the Arctic. The Antarctic sea-ice offers seasonal habitat hetereogeneity, as most of it melts in austral summer (Dayton et al. 1994). The Arctic sea ice is less dynamic and never melts except at the periphery. However, this may change rapidly within the next decades as the Arctic sea-ice shows substantial decreases in both extent and thickness in response to global warming. The Arctic region is less well defined by environmental characteristics, consisting of the Arctic Ocean with in the center a permanent cover of slowly circulating ice floes surrounded by a zone of seasonal pack-ice and a zone of land fast-ice (e.g. Stonehouse 1989). Polynias, predictable areas of open water within ice-covered seas, are of particular importance in Arctic habitats as they represent areas with high nutrient levels and enhanced productivity (Comiso & Gordon 1987; Dayton et al. 1994). Of all the Arctic pinniped species, only three have a continuous circumpolar distribution (ringed (*Phoca hispida*) and bearded seal (*Erignathus barbatus*) and walrus (*Odobenus rosmarus*) e.g. Stirling et al. 1983). However, several other species associate with the ice seasonally and are dependent on it as a breeding substrate. The Arctic pinniped species covered in this review are: ringed, bearded, harp (*Phoca groenlandica*), hooded (*Cystophora cristata*), grey (*Halichoerus grypus*), ribbon (*Phoca fasciata*), Caspian (*Phoca caspica*), Baikal (*Phoca siberica*), Largha (*Phoca largha*), harbour seal (*Phoca vitulina*) and walrus.

Seasonal or year-round association with ice offers pinnipeds a number of advantages, such as abundant food supply that is readily accessible under the ice in relative absence of competitors and terrestrial predators, a solid substrate to moult, rest, give birth and nurse pups. Dependent on the ice type, ice may also provide shelter in ridges and crevices as well as a milder micro-climate as there is generally less wind and wave action within ice packs (Riedman 1990).

PINNIPED ADAPTATIONS FOR LIFE ON ICE

Apart from a number of physiological adaptations for life in ice-dominated environments, such as sharp and strong claws for locomotion on the ice, the lanugo fur of pups which maintains body heat and thick subcutaneous blubber layers to restrict heat loss in adult seals (e.g. Lydersen & Kovacs 1999), polar pinnipeds also developed behavioural adaptations to life in their habitat. In temperate regions, the onset of parturition in terrestrial breeding pinnipeds is to a large extent determined by ambient temperature. Polar pinnipeds, however, depend on ice for breeding. Consequently, parturition does not occur in late spring and early summer, but rather in the late winter and early spring, when snow accumulation is at a maximum and temperatures are well below freezing (Pierotti & Pierotti 1980). This is the time of year when the ice is most extensive and stable and pup mortality as a result of ice breakup is minimized (Pierotti & Pierotti 1980). As this period of optimal ice conditions is relatively short, pupping is generally synchronous within ice-breeding pinniped populations compared to pinnipeds breeding on land. Grey seals provide a unique illustration in this respect as some grey seal populations breed on ice, whereas others breed on land. The grey seal populations that breed on ice have much more condensed lactation and birthing periods than the populations that breed on land, which is thought to be a response to the higher risk of premature separation of mother and pup on ice (Pierotti & Pierotti 1980; Haller et al. 1996). Nevertheless, within the ice habitats of polar pinnipeds there also exists considerable variation which appears to be of influence on behavioural patterns and the timing of behaviour as well (Trillmich 1996; Lydersen & Kovacs 1999). Variability in e.g. the temporal and structural stability of the platform, risk of predation, availability of food within the breeding habitats and access to the water are factors that have been suggested to have resulted in the evolution of different maternal strategies and consequently in differences in development of mother-pup acoustic communication in ice-breeding phocids (Insley 1992; Trillmich 1996; Lydersen & Kovacs 1999). In the light of current trends in climate change, knowledge on small scale local adaptations of behaviour in ice-breeding pinnipeds is of great importance in order to understand changes in abundance, distribution and behaviour. In the following section we provide an overview of the current state of knowledge on the role of in-air acoustic cues in polar pinniped mother-pup pairs and relate this to maternal strategies and breeding habitat characteristics.

ACOUSTIC BEHAVIOUR OF POLAR PINNIPED MOTHER-PUP PAIRS

In all pinniped species studied to date, pups have been found to vocalize in a similar fashion when interacting with the mother (e.g. Perry & Renouf 1988; McCulloch et al. 1999; Van Opzeeland & Van Parijs 2004; Collins et al. 2006). Individual stereotypy in calls is an important aspect of vocal recognition, as it enables individuals to distinguish between one and another, although individual vocal stereotypy does not necessarily indicate individual recognition (Insley 1992; Insley et al. 2003). Vocal signalling has been shown to play an important role in successful otariid mother-pup reunions upon the female's return from regular foraging trips (Trillmich 1981; Gisiner & Schusterman 1991; Insley 2001; Charrier et al. 2002). In phocids, however, the role of vocal signals in the recognition process, has been found to be more variable and has to date only been investigated in few species (see Table 1; Renouf 1984; Insley 1992; Job et al. 1995; McCulloch et al. 1999; McCulloch & Boness 2000; Van Opzeeland & Van Parijs 2004; Collins et al. 2005; 2006). Within ice-breeding phocids, evidence for individual stereotypy in pup calls has been found in the three colonial breeding species: Weddell seals (Collins et al. 2005; 2006), grey seals (McCulloch et al. 1999; McCulloch & Boness 2000) and harp seals (Van Opzeeland & Van Parijs 2004). However, the patterns in pup call individual stereotypy reported in these species are markedly different and are likely to reflect the complex interactions between e.g. the degree of coloniality, the likelihood and predictability of separations due to maternal foraging or ice break-up and the ontogeny of acoustic behaviour in ice-breeding pinnipeds (McCulloch & Bonness 2000; Insley et al. 2003; Van Opzeeland & Van Parijs 2004). In Weddell seals, females form breeding aggregations in fast ice areas where ice cracks provide access to the water. Compared to other phocid species, the lactation period in Weddell seals is relatively long, lasting 6-7 weeks (Laws 1981; 1984). During the first two weeks post-partum, females attend their pup on the ice continuously. However, during the second half of the nursing period, females spend increasingly more time in the water (Tedman & Bryden 1979; Hindell et al. 2002; Sato et al. 2003). Pups also start entering the water around this time, although the age at which pups first enter the water varies and is thought to depend on differences in local ice conditions and colony density (Tedman & Bryden 1979). Weddell seal pups vocalize and their vocalizations have been found to be moderately individually distinctive (Collins 2006). Unlike many other phocid mothers, Weddell seal females vocalize frequently to their pups (Kaufman et al. 1975; Collins et al. 2005). Female in-air calls have also been investigated in this species and have been found to exhibit individual stereotypy, although vocalizations are not unique (Collins et al. 2005). The critical amount of distinct information in both female and pup calls combined with an individual's visual and olfactory cues is likely to allow mother-pup pairs to recognize each other (Collins et al. 2005; 2006).

In grey seals, vocal behaviour of mother-pup pairs has to date only been studied in land-breeding populations. This has led to the finding of some remarkable differences between land-breeding colonies, which have been related to the ice-breeding ancestry of the species (McCulloch et al. 1999; McCulloch & Boness 2000). Through playback experiments, McCulloch et al. (1999; McCulloch & Boness 2000) were able to show that in the Sable Island colony, grey seal females discriminate between the vocalizations of their own and unfamiliar pups, whereas this ability appeared absent in the Isle of May grey seal colony. It was suggested that the female's ability to recognize the vocalizations of their own pup in the Sable Island colony could be a vestige of the ice-breeding ancestry of that colony, where it might have evolved in response to the higher risks of mother-pup separations (Pierotti & Pierotti 1980; McCulloch & Boness 2000). Acoustic behaviour of grey seal mother-pup pairs has, however, to date not been studied in any of the ice-breeding grey seal populations. As grey seals breed both on fast-ice and pack-ice, comparisons between these populations might provide interesting insights into the impact of breeding substrate stability on mother-pup acoustic behaviour.

Harp seal females form large breeding aggregations on seasonal Arctic pack-ice. During the 12 day lactation period, females forage a few hours per day (Lydersen & Kovacs 1993; Kovacs 1987; 1995), leaving their pup alone on the ice. Pups are relatively sedentary, rarely leaving the ice flow or entering the water. Harp seal pups vocalize in air during the lactation period, their vocalizations being structurally complex and variable (Miller & Murray 1995). Pup vocalizations were found to exhibit a relatively low percentage of individual variation (Van Opzeeland & Van Parijs 2004; Van Opzeeland et al. in prep). In the Greenland Sea population, vocalizations of female pups however, were found to be significantly more individually stereotyped within individuals than males, biasing maternal recognition towards female pups. However, in the Canadian Front harp seal population, no significant difference in pup vocal individuality between the sexes was found (Van Opzeeland et al. in prep). These differences in vocal individuality between the sexes may reflect different selection pressures working on female and male harp seals (Van Opzeeland & Van Parijs 2004). Alternatively, these population differences may be related to small scale local adaptations to i.e. site-specific ice conditions (Van Opzeeland et al. in prep). Clearly, further study is needed to investigate what is driving these differences in harp seal pup vocal behaviour.

Table 1. Review of current knowledge on polar pinnipeds concerning whelping habitat, gregariousness, duration of the lactation period, female foraging during lactation and the type of inidividualistic vocalization and recognition tested

Species		Whelping habitat	Gregarious	Duration of lactation (days)	Females at sea during lactation	Individualistic vocalization tested	Type of recognition tested
Harp seal		Pack-ice	Yes	12	Yes	Pup calls	None
Grey seal		Pack-ice, fast ice and land	No, No, Yes	12-17	Yes	Pup calls	Pup by mother
Harbour seal	Arctic	Pack-ice and land	Yes, Yes	24-42	Yes	Pup calls	Pup by mother
Hooded seal		Pack-ice	No	3-4	No	None	None
Bearded seal		Pack-ice	No	24	Yes	None	None
Ringed seal		Fast-ice	No	39-41	Yes	None	None
Largha seal		Pack-ice	No	14-21	No data	None	None
Caspian seal		Fast-ice	No	20-25	No data	None	None
Baikal seal		Fast-ice	No	60-75	Yes	None	None
Ribbon seal		Pack-ice	No	21-28	No data	None	None
Walrus		Pack-ice, Fast-ice	Yes	~730	Yes	Pup calls	Pup by mother
Weddell seal	Antarctic	Fast-ice	Yes	33-53	Yes	Mother + Pup calls	None
Crabeater seal		Pack-ice	No	17-28	No	None	None
Leopard seal		Pack-ice	No	~30	No data	None	None
Ross seal		Pack-ice	No	~30	No data	None	None

Table 2. Overview of passive acoustic techniques that are currently used to study marine mammals and their suitability to study polar pinnipeds (partially based on information derived from Van Parijs et al. 2007)

Technique	Duration deployment	(near) real-time data	Vessel requirements	Data storage capacity	Suitable for use on pinnipeds in polar areas	Available types
Ship-towed arrays	Hours to weeks	YES	Dedicated ship time	Essentially unlimited	Dependent on ice conditions	Ecologic Ltd., MAPS, many more
Acoustic tags	Hours to days	NO	Deployment and retrieval	A few gigabytes	YES	Bprobe, DTAG
Moored autonomous hydrophones (bottom, deep-sea mooring, ice based)	up to several years (dependent on sampling regime)	NO	Yearly or bi-yearly deployment and retrieval (but dependent on sampling regime)	Giga- to terabytes	YES (iceberg shifting in deeper waters)	Popup, HARP, ARP, AURAL-M2, EARS
Gliders and underwater vehicles	Weeks to months	YES	Deployment and retrieval	Gigabytes	YES	SeaGlider, WHOI
Radio-linked sonobuoys	Hours to months	YES	Dedicated ship/air time	Essentially unlimted	YES	Military surplus, DIFAR
Cabled systems	Years to decades	YES	One-time deployment; maintenance	Essentially unlimited	NO	SOSUS, ALOHA, NEPTUNE, AUTEC, CTBTO
Autonomous listening stations	Years to decades	YES	One-time deployment; maintenance	Essentially unlimited	YES	PALAOA

Although most harbour seal populations form dispersed breeding colonies on land, few also give birth on ice (Streveler 1979; Calambokidis et al. 1987). Research on these populations is limited and to date no study has addressed mother-pup behaviour in ice-breeding harbour seals. In land-breeding harbour seals, pups often accompany their mothers on foraging trips from birth (Wilson 1974). Pups vocalize both in-air and in water and although airborne and underwater vocalizations were found to differ, pup calls were individually stereotyped (Renouf 1984; Perry & Renouf 1988). However, to what extent these findings can be extrapolated to ice-breeding harbour seals is unknown.

For solitary ice-breeding pinnipeds the selective pressures favouring development of individually stereotyped vocalizations may be less strong as there is little confusion possible over maternal investment. Crabeater and hooded seals are solitary pack-ice breeders and have short lactation periods, during which females remain with their pup on the ice throughout the nursing period (Siniff et al. 1979; Riedman 1990). Consequently, there is little opportunity for mother-pup pairs to become separated. Hooded seal pups emit snorts, grunts or brief low-frequency moans while attended by their mothers. Ballard & Kovacs (1995) concluded that these vocalizations are unlikely to be used by a female to identify her pup as these sounds contain little frequency or amplitude modulation which in many other species have been found to bear the individually distinctive cues (e.g. Phillips & Stirling 2000; Charrier et al. 2002). In crabeater seals, nothing is known on the role of vocal behaviour in mother-pup interactions, although vocalizing has been reported to occur when the pair is separated (Siniff et al. 1979). In bearded and ringed seals, both pup and female are known to forage throughout the lactation period (Hammill et al. 1991; Lydersen & Hammill 1993; Hammill et al. 1994; Kelly & Wartzok 1996; Krafft et al. 2000). Vocal cues might therefore serve a function to coordinate and synchronize mother-pup behaviour during the lactation period. However, to date the role of acoustics in mother-pup interactions in these species has not been investigated.

In the other pack-ice breeding pinnipeds, Ross, leopard, Largha, Caspian, Baikal and ribbon seals, knowledge on the species' general biology is to a large extend still lacking and nothing is known on acoustic behaviour in mother-pup pairs.

Walruses also breed on pack-ice, forming dense aggregations in spring. Calves enter the water immediately after birth and are nursed for at least one year both on the ice and in the water (Riedman 1990; Kastelein 2002). However, most calves associate with their mothers in groups of adult females for longer periods and are weaned after three years (Kastelein 2002). Walrus female-offspring acoustic recognition has been suggested by observation to be well developed in walruses (Kibal'chich & Lisitsina 1979; Miller & Boness 1983; Miller 1985; Kastelein et al. 1995) and was recently also experimentally demonstrated. Walrus calf vocalizations were found to be highly stereotyped and females were found to respond more strongly to playbacks of vocalizations of their own calf than to the calls of an alien calf (Charrier pers. comm.).

ACOUSTIC BEHAVIOUR AND MATING STRATEGIES IN POLAR PINNIPEDS

The transition from ancestral terrestrial parturition to giving birth on ice is also thought to have had major consequences for the evolution of mating strategies in ice-breeding pinniped species (e.g. Bartholomew 1970; Pierotti & Pierotti 1980; LeBoeuf 1986; Van Parijs 2003). On land, the relative rarity of suitable pupping and haul-out areas causes the formation of very dense female breeding aggregations, enabling males to defend harems or compete with other males for a place within a female breeding group (Bartholomew 1970). However, ice habitats generally offer large areas suitable for parturition and haul-out and consequently many ice-breeding pinnipeds aggregate in loose colonies or breed in a solitary fashion (e.g. Stirling 1975; LeBoeuf 1986; Lydersen & Kovacs 1999). This, along with the fact that females in many ice-breeding pinnipeds forage to sustain lactation, causes female movements to be both spatially and temporally less predictable for males compared to land-breeding females (Van Parijs 2003). As a consequence, ice-breeding pinniped females cannot be economically monopolized by males when they become receptive and male reproductive strategies must aim to attract females for the purpose of mating (Van Parijs 2003). All ice-breeding pinnipeds mate aquatically and underwater vocalizations and stereotypical dive displays are known to form an important part of male-male competition and male advertisement to females in aquatic mating species (see Van Parijs 2003 for a review). The available evidence appears to indicate the existence of different mating systems within aquatic mating species (Kovacs 1990; Rogers et al. 1996; Van Parijs et al. 1997; 2001; 2003; Harcourt et al. 2007; in press). However, due to the difficulties of studying ice-breeding pinnipeds, too few species have been studied to date to compare the relative impact of habitat and female behaviour on male mating tactics. Here we summarize what is currently known on acoustic behaviour related to mating behaviour in ice-breeding pinnipeds.

In colonial breeding species such as Weddell and harp seals, communication generally occurs over relatively short distances as both males and females form seasonal aggregations. Signals are not constrained by propagation needs and consequently many different sound types as well as subtle variations in sounds are used in communication. Accordingly, the vocal repertoires of colonial breeding species are generally broad and consist of a wide variety of sounds that serve local advertisement displays in order to defend territories and to attract mates (Rogers 2003). Male Weddell seals typically defend underwater territories around or near tide cracks used by females, perform short shallow display dives and have a large underwater vocal repertoire, including the male-specific long descending "trill" (Kooyman 1981; Thomas & Kuechle 1982; Bartsch et al. 1992; Oetelaar et al. 2003; Harcourt et al. in press). Vocal activity increases strongly during the breeding season (Green & Burton 1988; Rouget et al. 2007) and trill vocalizations are likely used underwater by males for the purpose of territorial defence, advertisement, dominance and warning signals (Thomas & Kuechle 1982; Thomas & Stirling 1983; Thomas et al. 1983). Although female movements are somewhat predictable as females use tide cracks in the ice to access the water, females have access to all parts of the water column and male monopolization of females may be difficult (Hindell et al. 2002; Sato et al. 2003; Harcourt et al. 2007; in press). It has been suggested that in systems where males cannot monopolize females, male-male competition may play a less important role (e.g. Harcourt et al. 2007). Male territories under the ice may instead primarily serve to maximize exposure of the territory holder to females passing through. Male vocal and dive displays may be used by females to assess quality of a potential mating partner and consequently female choice may have a significant role in mating success. However, male Weddell seal mating tactics have also been found to exhibit plasticity (Harcourt et al. in press) and more detailed investigation of underwater behaviour of male and female Weddell seals is needed to test this hypothesis.

Similar to Weddell seals, harp seals also have a large vocal repertoire consisting of a wide variety of sounds that suit local communication purposes (Møhl et al. 1975; Terhune & Ronald 1986; Terhune 1994; Serrano 2001; Rogers 2003). Harp seal vocal activity increases in both sexes during the breeding season, suggesting that females also may have an important role in mating behaviour (Watkins & Schevill 1979; Serrano & Miller 2000; Serrano 2001). Harp seal females aggregate in colonies and use leads between shifting pack-ice floes to access the water to forage during lactation. Males therefore have access to clusters of females. Merdsøy et al. (1978) reported large male harp seal groups travelling through the breeding herd early in the breeding season. Agonistic interactions between males increased towards the time that pups were weaned and males and females were seen hauled out on the ice (Merdsøy et al. 1978). Males have been observed snorting and bubble blowing at holes used by females (Merdsøy et al. 1978; Kovacs 1995). However, to date it is unknown how vocal behaviour relates to harp seal male and female mating behaviour.

Ringed seals also exhibit a relatively rich vocal repertoire, which is thought to serve the purpose of male local display (Stirling 1973; Kunnasranta et al. 1996; Rogers 2003). Although ringed seals do not breed in colonies, they often form small aggregations on fast-ice. Females are believed to maintain birth lair complexes which are included in an area occupied by a territorial male (Smith & Hammill 1981). Interestingly, Weddell, harp and ringed seals have been found to vocalize year round, with peaks in vocal activity during the breeding season (Green & Burton 1988; Kunnasranta et al. 1996; Serrano 2001; Rouget et al. 2007). Apart from the sole purpose of vocal display during the mating season, vocal behaviour has in these species been suggested to also serve other purposes such as social communicative function during migration or pursuit of prey (Kunnasranta et al. 1996; Serrano & Miller 2000; Rouget et al. 2007). However, only few studies have investigated the vocal behaviour of these species outside the breeding season (Serrano & Miller 2000; Serrano 2001; Rouget et al. 2007)

In contrast to the rich vocal repertoire of Weddell, harp and ringed seals, a number of polar pinnipeds produce single or series of relatively short broadband pulsed sounds, which have been suggested to mainly function in agonistic interactions (Rogers 2003; Hayes et al. 2004). Land-breeding harbour seals perform short dives and produce underwater roar vocalizations in underwater display areas (Hanggi & Schusterman 1994; Van Parijs et al. 1997; Hayes et al. 2004). Male mating strategies were found to be closely linked to habitat type and resulting changes in female behaviour, distribution and density (Van Parijs et al. 1999; 2000). In ice-breeding harbour seals, females have been reported to strongly depend on the limited availability of suitable haul-out ice (Calambokidis et al. 1987; Mathews & Kelly 1996). Similar to land-breeding harbour seals, this may enable males to concentrate and display in areas that are frequented by females. However, the underwater vocal behaviour in ice-breeding harbour seals has not been studied and it is not known if males in these populations also hold underwater territories. Grey seals also breed both on ice and on land and have a similarly simple vocal repertoire consisting of short gutteral sounds, growls and clicks (Schusterman et al. 1970; Asselin et al. 1993). Breeding habitat has in this species been shown to be of great influence on mating behaviour (Anderson & Harwood 1985). Whereas males in land-breeding grey seals defend harems, males in ice-breeding grey seals are unable to monopolize females that are widely dispersed on the ice. Males are usually seen attending one female and her pup on the ice, forming triads (Riedman 1990). Another difference between ice-breeding and land-breeding grey seals is the fact that ice-breeding grey seal males and females regularly enter the water during the breeding season, whereas this is rarely seen in land-breeding populations (Asselin et al. 1993). The underwater vocal repertoire of grey seals has also been found to differ between land- and ice-breeding populations (Asselin et al. 1993). Underwater vocal display may therefore form an important part of ice-breeding grey seal male mating behaviour.

For crabeater seals, only one short broadband call type has been documented and vocal activity is thought to be restricted to the breeding season (Stirling & Siniff 1979; Rogers 2003). Crabeater seals form triads and occur in low densities in pack-ice areas. Males attend one female and her pup on the ice and defend the female against intrusions by other adult crabeater seal males until the female becomes receptive (Siniff et al. 1979). The relatively simple acoustic display of crabeater seal males is thought to function primarily in short-range male-male competition in guarding the female, as loud complex vocalizations would have the potential to attract other distant males to the pre-oestrus female (Rogers 2003).

Male hooded seals have a small in-air and underwater acoustic repertoire, involving five call types, most of which were found to be used in close-range communication in agonistic or sexual contexts during the reproductive period (Ballard & Kovacs 1995). Similar to crabeater seals, hooded seals form triads on the ice. However, hooded seal females are generall less widely dispersed compared to crabeater seals (e.g. Sergeant 1974; Siniff et al. 1979; Boness et al. 1988). Males may therefore move more easily between females resulting in some degree of polygyny (Boness et al.1988; Kovacs 1990). In addition, observations of some males attending a female continuously on the ice, whereas others were more mobile and attended several females for shorter periods of time, are suggestive of the use of alternative mating strategies by hooded seal males (Kovacs 1990). Visual displays involving the hood and septum form an important part of male hooded seal behaviour as male displays to a large part take place on the ice (Kovacs 1990; Ballard & Kovacs 1995).

In walruses, males of Atlantic and Pacific populations have been found to use different mating tactics. In the Atlantic walrus, large mature males were observed to attend and monopolize groups of potentially reproductive females for extended periods (Sjare & Stirling 1996). Male distribution in this population was mainly determined by ice-cracks and polynias that provided easy access to open water (Sjare & Stirling 1996). Pacific walruses breed on drifting pack-ice with a rapidly changing distribution of open water and much higher breeding population densities (Fay et al. 1984). The highly unstable environment combined with higher densities of potentially reproductive females is thought to make it more advantageous for Pacific walrus males to display in small areas for brief periods than to continuously attend and defend one herd (Sjare & Stirling 1996). Male walruses vocalize extensively in the vicinity of females and calves, emitting short repetitious pulses which have been suggested to exhibit individual stereotypy (Stirling et al. 1987).

In solitary pack-ice or fast-ice breeders, individuals need to broadcast their sounds over long distances to advertise their position to potential mates and rival males (Van Parijs 2003; Rogers 2003). These species generally have a medium size repertoire and vocalizations tend to be stereotyped signals to increase the likelihood that they are received by the intended recipient (Rogers 2003). Only limited recordings have been made of ribbon seal vocalizations. Watkins and Ray (1977) recorded ribbon seal high amplitude downward sweeps and puffing sounds towards the end of the breeding season. The sounds are thought to be produced by males, as only males were found to have well-developed air sacs, which may aid in the production of loud underwater sounds (Stirling & Thomas 2003). However, to date nothing is known on the role of vocal behaviour in ribbon seal mating behaviour. No study has investigated the behaviour of Caspian and Baikal seals and no vocalizations have been recorded.

Leopard seal male and females are widely dispersed at the start of the mating season as females give birth, nurse and wean their pups alone on the pack-ice. Communication must occur over long distances and both females and males have been found to produce loud broadcast calls during the breeding season (Rogers et al. 1996; Rogers & Bryden 1997). Lone males are known to vocalize for many hours each day, which may serve as an indicator of male fitness as these displays require the male to be in good body condition (Rogers 2003; 2007). Females are thought to produce broadcast calls to advertise their sexual receptivity to distant males (Rogers et al. 1996). The use of long-distance broadcast calls has also been suggested to occur in bearded and Ross seals (Rogers et al. 1996). Little is known on Ross seal mating behaviour as the species occurs in low densities in heavy pack-ice areas. Vocalizations have only been recorded in December through January, which suggests that these vocalizations are related to the mating season (Watkins & Ray 1985; Stacey et al. 2006; Seibert 2007). Ross seal vocalizations have been described as 'siren calls' (Watkins & Ray 1985) and are loud and semi-continuous, which makes them suitable to communicate over long distances (Rogers 2003).

Male bearded seals use loud trilling vocalizations which have been found to carry over large distances to advertise their breeding condition to females (Cleator et al. 1989; Van Parijs et al. 2003). Females are dispersed, but their movements are largely restricted to areas with suitable haul-out ice (Burns 1981). During the breeding season, male bearded seals have been found to vocalize in higher densities in areas where oestrus females are found regularly (Van Parijs et al. 2001; 2003). In Svalbard, bearded seals males have been found to use alternative mating tactics, where some males 'roam', displaying over large areas, whereas others are territorial and display over smaller areas (Van Parijs et al. 2003). Territorial males had longer trills than roaming males, which may be used by females as an indicator of male quality (Van Parijs et al. 2003). In addition, male mating success was shown to be dependent on variation in breeding habitat as increased ice cover was found to restrict the number of roaming males, whereas territorial males were present during all ice conditions (Van Parijs et al. 2004). In the light of current climatic trends, changes in ice-associated habitat may therefore alter the long-term mating success of individual male bearded seals. Although predictions on the potential effects of climate change on polar pinnipeds mainly concern regional or seasonal shifts in prey availability and changes in timing and patterns of migration (Tynan & DeMaster 1997; Friedlaender et al. 2007), small scale behavioural changes should not be ignored as important indicators of change. Acoustic techniques are a useful tool to study pinniped behaviour and may therefore also provide important insights in the potential effects of climate change on polar pinnipeds.

POLAR PINNIPEDS AND ANTHROPOGENIC NOISE

Rapidly increasing anthropogenic noise levels in the ocean and the impact of this noise on marine mammals have become a growing concern over the last years. With respect to polar habitats, ice breaker vessels and shipping traffic form the predominant anthropogenic noise sources to which ice-breeding pinnipeds are exposed. Recent computer projections of the National Snow and Ice Data Center have indicated that the receding Arctic sea-ice will leave more and more areas partially or largely ice-free year-round within the near future. Consequently, this opens opportunities to re-route commercial vessel traffic to and from Asia to take advantage of the open Northwest Passage. The increased shipping activity and the year-round presence of vessels in these areas will lead to substantially increased noise levels. These changes are likely to have consequences for polar pinnipeds that aggregate to mate and give birth to pups in traditional areas within these regions. Evidence of the potential impact of vessel sounds on pinniped behaviour and acoustic communication is generally meager (Richardson et al. 1995). In harp seals, calling rates were found to decrease after vessels came within 2km of the whelping area (Terhune et al. 1979). It was uncertain if calling rates decreased because animals stopped vocalizing or because they left the area. Further study is clearly needed and should include the acoustic monitoring of the areas of anticipated increases in vessel activity and the acoustic behaviour of ice-breeding pinnipeds within these regions.

In the next section we provide an overview of new and emerging passive acoustic technologies that can be used to further the study of polar pinnipeds.

ACOUSTIC DATA COLLECTION

Acoustic techniques only recently entered the range of easily accessible research tools, as significant advances in audio and computer technology now allow the acquisition and handling of large acoustic data sets. As a consequence, acoustic techniques have become increasingly important as a tool for remote sensing behaviour of various marine mammal species (e.g. Stafford et al. 1998; McDonald & Fox 1999; Janik 2000; Johnson & Tyack 2003; Mellinger et al. 2007). Compared to visual observation, acoustic recordings are quasi-omnidirectional and independent of light and weather conditions, providing the option of detecting and studying animals at night and under conditions where visual observation are not possible (see Erbe 2000 for a comparative discussion of acoustic and visual censuses). In particular for species in offshore or remote polar areas, newly developed acoustic techniques allow investigation of these animals in their natural habitat for extended time periods. However, not all techniques are equally well suited for collecting data in polar areas, as ice cover and harsh weather conditions can frequently limit deployment. In addition, instrumentation features such as the possibility of recording over longer time spans, the need for a vessel or on-site operators and access to real time data, determine which type of acoustic instrumentation is most suitable for specific research purposes. A comprehensive review of acoustic observation methods can be found in Mellinger et al. (2007). Here we provide a brief overview of the types of passive acoustic techniques that are currently used to study marine mammals. However, given the scope of this review, devices and techniques will be discussed in the light of their suitability for studying pinnipeds in polar environments.

ACOUSTIC INSTRUMENTATION

Ship-towed *hydrophone arrays* allowing coverage of relatively large areas, are in most cases relatively cost-efficient and can be combined with visual surveys (e.g. Spikes & Clark 1996; Norris et al. 1999). Towed systems can be deployed in offshore or remote areas, but often require dedicated ship time and personnel. The time-spans during which towed arrays are deployed are therefore generally relatively short (i.e. hours to weeks). Successful use in polar environments is largely dependent on ice conditions as heavy ice cover may limit access by ships and may damage recording gear. The use of towed arrays in ice-covered areas has nevertheless been shown to be feasible (e.g. Kindermann et al. 2006). Of greater concern are the high noise levels generated by icebreaker vessels which are likely to mask the majority of animal vocalizations, particularly in the mid (< 10kHz) to low (< 1kHz) frequency ranges. However, if the array consists of 10 or more tightly spaced hydrophones, beamforming techniques can be used to significantly improve signal-to-noise-ratios (e.g. Mellinger et al. 2007).

Acoustic tags (e.g. DTAG, Bioacoustic Probe) are miniature acoustic recorders that can be attached to marine animals to collect data on the acoustic stimuli emitted and experienced by the tagged subject (e.g. Johnson & Tyack 2003). Additionally, these devices are also capable of sampling various environmental variables as well as physiological and behavioural data (Fletcher et al. 1996; Madsen et al. 2002). The use of acoustic tags can provide particularly useful information about an individual's behaviour in relation to its vocalizations and sounds from its environment. This technique is also used as a tool to determine correction factors for marine mammal surveys such as the amount of time animals are vocalizing (e.g. Erbe, 2000). However, when animals are in close groups it can be difficult to separate whether the vocalization is produced by the tagged individual or one nearby. Furthermore, acoustic tags can only sample data over periods of hours to days. In contrast to satellite telemetry tags which transmit data back to a receiving station, acoustic tags need to be retrieved before data can be analysed, which might be a difficulty if tags come off in areas that are not easily accessed (i.e. ice covered regions). In addition, as is the case with all tagging studies, the influence of both the tagging event and the presence of the tag on the individual need to be ascertained before major conclusions are drawn from such data.

Autonomous recording devices consist of a hydrophone and a battery-powered data recording system. These instruments are either free-drifting (surface recording units, e.g. Hayes et al. 2000; Collison & Dosso 2003), moored on the sea floor or attached to deep-sea moorings (e.g. Calupca et al. 2000; Newcomb et al. 2002, Wiggins 2003). Alternatively, these devices can be ice-based with hydrophones deployed through holes in the ice (Klinck 2008). Ice-based autonomous recording devices also offer the possibility for in-air recording of vocalizing animals that are hauled out on the ice (e.g. Collins et al. 2005; 2006). These devices are battery-powered and record and store acoustic data internally. Dependent on data storage capacity of the device, recording bandwidth and sampling regime, recordings can be obtained over extended periods of time, in some cases up to several years (e.g. Wiggins 2003;

Sirovic et al. 2003; Moore et al. 2006). In addition, by deploying several synchronized devices in an array, large areas can be acoustically monitored and movements of animals within these areas can be tracked. However, only archival data collection is possible and consequently, data analysis can only occur after a certain period of time when devices have been recovered. Autonomous recording devices have the advantage that they can be deployed in a wide variety of areas, including polar environments (e.g. Wiggens 2003; Moore et al. 2006). Nevertheless, in ice-covered areas, deployment may be restricted to areas with greater water depths (i.e. >250m in Antarctic regions) as in shallow waters, drifting icebergs often cause damage to moored instruments.

An emerging passive acoustic technique involves the use of *autonomous underwater vehicles and gliders*, which originally were developed for oceanographic research (e.g. Eriksen et al. 2001; Sherman et al. 2001). More recently these devices have been used for various research purposes and have been successfully deployed in ice-covered areas (e.g. Brierley et al. 2002; Owens 2006). Gliders move vertically and horizontally and can be remotely controlled at regular time intervals through an intrinsic two-way communication system, which also allows stored data to be transmitted back to the lab for immediate analysis. However, the type of bi-directional satellite transmitters that are currently used in gliders allow transmission of limited amounts of data only (see also the discussion on satellite transmission below). Furthermore, gliders require significant resources to build and maintain. In cases where they are deployed for acoustic recording purposes, devices need to be recovered after weeks or months to retrieve data. With respect to polar areas, gliders could nevertheless provide a tool for remotely controlled acoustic sampling of (periodically) inaccessible areas. Bioacoustic research using these devices is nevertheless still limited (Baumgartner et al. 2006; Fucile et al. 2006).

Free-drifting *radio-linked sonobuoys* enable short term real time acoustic monitoring as they transmit acoustic recordings as a radio signal and are often deployed from vessels or aircraft. In order to transmit recordings in good quality, radio signals require a receiver to be in relative proximity of the recording device, which makes these systems suitable for use during ship-based surveys where visual and acoustic observations are combined (e.g. Clark et al. 1994; Rankin et al. 2005). Some drifting radio-linked sonobuoy types also allow localization by giving a compass bearing to the sound source (DIFAR buoys, e.g. Greene et al. 2004). The duration of the period over which these devices transmit acoustic recordings and the cost of devices strongly depends on the type that is used. With respect to polar areas, successful use of these devices will largely depend on ice coverage as shifting ice may cause damage or block transmission of the radio signal. In heavy ice-covered areas, radio-linked sonobuoys can be fixed on the ice surface while the hydrophone is deployed through a borehole or crack in the ice (e.g. Clark et al. 1996).

Cabled passive acoustic recording stations can be operated continuously in offshore areas without limitations on data storage capacity and power supply. In addition, cabled recording stations allow near-real time monitoring which enables the linking of acoustic recordings to visual observations. However, the majority of cabled arrays in offshore areas require significant resources and are predominately operated by government institutions. An example of such a station is the US Navy sound surveillance system (SOSUS array), which consists of a network of hydrophones covering the deep offshore waters throughout the Atlantic and Pacific oceans. Due to the military purposes of these systems, access to the acoustic data is often restricted, only off-line available and the recording bandwidth of these systems is often limited to lower frequencies. These systems have nevertheless been used successfully to study baleen whales and other species calling at low frequencies (e.g. Clark 1995; Clark & Charif 1998; Moore et al. 1998; Stafford et al 1998; 1999).

In recent years, however, a growing number of non-military cabled acoustic observatories are operated in coastal areas (e.g. ALOHA, Petitt et al. 2002; NEPTUNE, Barnes et al. 2007; AUTEC). These systems record continuously over broad frequency bandwidths, allow real time monitoring and localization of marine mammals and have no restrictions to data storage, data access and power supply. These characteristics would make cabled systems ideal if they could be used for data acquisition in remote polar environments. However, monitoring of polar regions with a network of hydrophones connected by cables to shore-based stations requires extreme cable length which would imply extensive deployment and maintenance costs. In addition, ice movements and cable melt-in can cause damage to long cables in polar environments. The use of cables can nevertheless be overcome by satellite or iridium phone mediated transfer of acoustic data directly from recordings units to receiver stations where data are analysed. As mentioned previously, the type of transmitter systems that are suitable to be integrated in recording units limit satellite data transmission rates, rendering them too low to allow continuous broadband acoustic data to be transmitted. This can be surmounted by only transmitting acoustic snippets or events (e.g. click detectors, TPODs; e.g. Tregenza 1999), which requires pre-processing of the signal within the buoy to detect and select events that are transmitted.

Transmission of continuous broadband acoustic data instead of snippets is possible if data transmission is mediated by a station with enlarged satellite receivers. Acoustic data from the recording unit can be transmitted to the satellite-linked station using a radio signal or a wireless local area network (WLAN) link, provided that the area is relatively flat and the satellite-linked station and the recording unit are not too far apart. The satellite-linked station can then be used to transmit acoustic data to receiver stations over large distances (an example of such a system will be discussed in the next section). Real time data transmission of autonomous recording units has the additional advantage that data does not necessarily have to be stored locally and that analyses can be performed near-real time (e.g. Simard et al. 2006).

In addition to data transfer, cables also serve to secure the continuous power supply of cabled systems. A system with satellite- and radio-linked data transmission is devoid of this option and therefore either dependent on batteries that need regular exchange or requires autonomous power supply. With respect to polar areas, autonomous power supply secures continuous powering of the station also when the area is periodically inaccessible.

PALAOA

The PerenniAL Acoustic Observatory in the Antarctic Ocean (PALAOA) is an example of a stationary autonomous listening station that records sound continuously year-round and provides online access to the real time data (see http://www.awi.de/acoustics). The PALAOA observatory is located at 70°31'S 8°13'W, on the Ekström Ice Shelf, Eastern Weddell Sea, at 1 km distance from the ice shelf edge (Figure 1). The main sensor was designed as a 520m baseline tetrahedral hydrophone array deployed through boreholes underneath the 100m thick floating Antarctic ice shelf (Boebel et al. 2006; Klinck 2008b)[1]. It is energetically self-sustained by utilizing solar and wind energy. A CTD probe is mounted to collect oceanographic readings while sea ice conditions of the adjacent ocean are monitored with a webcam. A 13km WLAN link connects the station to the German Antarctic Neumayer Base, which is manned year-round and has a leased satellite internet connection. Development efforts focused on the one hand on the real time transmission of a highly compressed live stream (24kbit/ogg-vorbis-coded) via a satellite link to the Alfred Wegener Institute for Polar and Marine Research (AWI) in Bremerhaven, Germany for immediate processing. On the other hand, acoustic data of very high quality (up to 4 channels, 192 kHz/24Bit uncompressed) is buffered on request at PALAOA and Neumayer Base respectively and can be downloaded for detailed analysis. PALAOA has been operational since December 2005 and has collected a total of 10000 hours audio so far (by January 2008). Recordings contain vocalizations of four Antarctic pinniped species (crabeater, Weddell, Ross and leopard seals) and a variety of cetacean vocalizations. In addition, the recordings contain sounds of abiotic origin, such as iceberg calvings and collisions. Current analyses aim to explore temporal and spatial distribution patterns of vocalizing individuals of the different species. Additionally, the PALAOA recordings are used to gauge the local ocean noise budget and monitor the impact of human activities on marine mammal behaviour. PALAOA provides an example of a state-of-the art system which allows data to be obtained for long term monitoring of acoustic underwater sounds in the Antarctic which has never been attempted previously.

SOFTWARE TECHNOLOGIES

As the data storage capacity of acoustic recording instruments has increased substantially over the last years, recordings can be made over longer time spans. In many cases these long term acoustic datasets require the use of automated detection and classification techniques, as manual detection and analysis becomes too time-consuming. A wide variety of software technologies are available to perform automated detection and classification (e.g. open source: Mellinger 2001; Figueroa 2006) and a number of different detection methods have been developed (e.g. see Mellinger et al. 2007 for a summary of methods). However, not all techniques are equally well fitted for different species, research goals and recording types. Species-specific vocal characteristics are one factor to consider when deciding for a specific software tool for analyses. Techniques involving matched filters or spectrogram correlation are most suitable to investigate species with stereotyped vocalizations, whereas more variable vocal patterns (e.g. dolphin whistles) are best detected using energy summation in specified frequency bands (e.g. Oswald et al. 2004). Also, the rarity of a species' vocalizations may determine the optimal configuration of the detector to achieve a trade-off between missed calls (false negatives) and incorrect detections (false positives). In species that are not very vocal or occur infrequently, the importance of detecting as many target vocalizations as possible may overcome the effort of an additional check of the detector's output (either manually or by using subsequent automated classifiers). Similarly, the purpose of the detection will also determine which is the optimal detector type and sensitivity (see Mellinger et al. 2007 for a discussion). A final factor to consider is the type of recordings. As has been

---

[1] Since mid 2006 the failure of two hydrophones has reduced the array to a two-channel system

mentioned earlier, multi-channel recordings can significantly improve signal-to-noise-ratios through beamforming and allow source localisation

Within marine mammal acoustic research, automated detection has to date almost exclusively been used to scan large data volumes for cetacean vocalizations (e.g. Stafford 1995; Mellinger & Clark 2000; Gillespie 2004; Lopatka et al. 2005; 2006). This likely reflects the fact that long term acoustic observations are much more embedded within cetacean research as compared to studies of pinniped behaviour. However, many of the basic research questions that still need to be addressed for a number of polar pinniped species require acoustic observation over longer time spans and consequently the use of automated detection techniques. Klinck et al. (2008; in press) recently applied various automated detection techniques to long term recordings of pinniped vocalizations and found, for example, that Hidden Markov Models, which are also used for human speech recognition (e.g. Juang & Rabiner 1991) perform well for leopard seal vocalizations. Nevertheless, as mentioned earlier, species-specific vocal behaviour and research goals will ultimately determine the suitability of an automated detection technique.

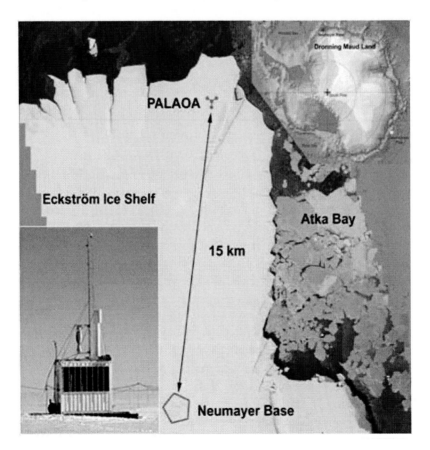

Figure 1. IKONOS-2 satellite image from March 2004, with locations of the Neumayer Base and PALAOA. Top inset: Antarctica with the location of PALAOA indicated by a red dot. Bottom inset: PALAOA Station.

CONCLUSIONS AND FUTURE RESEARCH DIRECTIONS

Little is known about the acoustics of most ice-breeding pinnipeds. Clearly, the inaccessible nature of their habitat plays an important role in explaining this short coming and has greatly influenced the extent to which these species have been studied. To date mother-pup vocal recognition has not been experimentally tested in any of the ice-breeding species, except for the walrus. Attempting to address these short comings would significantly increase our understanding of ice breeding species requirements in terms of offspring recognition. Investigations of vocal individuality of mother and pup calls show markedly different patterns for each species, which likely reflects the complexity of interactions that shape their vocal behaviour.

With respect to mating strategies, there is a distinction between the species that have been investigated in some detail, such as the bearded, Weddell seal and the walrus and those about which we known very little. Future investigations of well known species' mating systems require more small scale research focussing on individual differences and plasticity in male mating tactics. Additionally, the role of females and female choice also needs to be taken into account. For little known species an attempt at understanding the broader scale role of underwater vocalisations should be the initial focus. Recent changes in acoustic instrumentation technologies and their availability in terms of reduced costs should greatly facilitate the study of little known polar species. Although reseachers have primarily used novel acoustic instrumentation and software technologies for studying cetaceans, the majority of these devices and techniques provide vast opportunities for the study of polar pinnipeds.

It is critical to improve our understanding of both recognition processes and reproductive strategies of polar pinnipeds given current trends in climate driven changes which are altering their ice-dominated environments at hereto unprecedented rates. Ice-associated seals, which rely on suitable ice substrate for e.g. resting, pupping and moulting, are particularly vulnerable to climatic change. Similarly, changing conditions in the ocean basin such as increasing background noise, in terms of anthropogenic sounds, are becoming of heightened concern for polar environments. Among other techniques, acoustic techniques should be recognized as an extremely versatile and useful technology for future studies of pinniped ecology.

ACKNOWLEDGEMENTS

We thank Holger Klinck and Robert Huisman for providing constructive comments on this manuscript.

REFERENCES

Anderson, S. S. & Harwood, J. 1985. Time budgets and topography: how energy reserves and terrain determine the breeding behaviour of grey seals. *Animal Behaviour*, 33, 1343-1348.

Asselin, S., Hammill, M. O. & Barrette, C. 1993. Underwater vocalizations of ice-breeding grey seals. *Canadian Journal of Zoology*, 71, 2211-2219.

AUTEC. The Atlantic Undersea Test and Evaluation Centre. Retrieved (2008, January) from http://www.npt.nuwc.navy.mil/autec/home.htm.

Ballard, K. A. & Kovacs, K. M. 1995. The acoustic repertoire of hooded seals (*Cystophora cristata*). *Canadian Journal of Zoology*, 73, 1362-1374.

Barnes, C. R., Best, M. M. R., Bornhold, B. D., Juniper, S. K., Pirenne, B. & Phibbs, P. 2007. The NEPTUNE Project - a cabled ocean observatory in the NE Pacific: Overview, challenges and scientific objectives for the installation and operation of Stage I in Canadian Waters. *Symposium on Underwater Technology and Workshop on Scientific Use of Submarine Cables and Related Technologies*. Tokyo, Japan: IEEE Press, Piscataway, NJ.

Bartsh, S. S., Johnston, S. D. & Siniff, D. B. 1992. Territorial behaviour and breeding frequency of male Weddell seals (*Leptonychotes weddelli*) in relation to age, size and concentrations of serum testosterone and cortisol. *Canadian Journal of Fisheries and Aquatic Sciences*, 70, 680-692.

Batholomew, G. A. 1970. A model for the evolution of pinniped polygyny. *Evolution*, 24, 546-559.

Baumgarner, M. F., Fratantoni, D. M. & Clark, C. W. 2006. Investigating baleen whale ecology with simultaneous oceanographic and acoustic observations from autonomous underwater vehicles. *Eos Transactions, American Geophysical Union*, 87(36), Ocean Sciences Meeting Supplement. Honolulu, Hawaii. February 20-24, 2006. Abstract OS24E-05.

Boebel, O., Kindermann, L., Klinck, H., Bornemann, H., Ploetz, J., Steinhage, D., Riedel, S. & Burkhardt, E. 2006. Real-time underwater sounds from the Southern Ocean. *Eos Transactions, American Geophysical Union*, 87, 361-362.

Boness, D. J., Bowen, W. D. & Oftedal, O. T. 1988. Evidence of polygyny from spatial patterns in hooded seals (*Cystophora cristata*). *Canadian Journal of Zoology*, 66, 703-706.

Brierley, A. S., Fernandes, P. G., Brandon, M. A., Armstrong, F., Millard, N. W., McPhail, S. D., Stevenson, P., Pebody, P., Perrett, J. R., Squires, M., Bone, D. G. & Griffiths, G. 2002. Antarctic krill under sea ice: elevated abundance in a narrow band just south of ice edge. *Science*, 295, 1890-1892.

Burns, J. J. 1981. Bearded seal, *Erignathus barbatus* (Erxleben 1777). In: *Handbook of Marine Mammals* (Ed. by Harrison, S. H. R. R. J.), pp. 145-170. London: Academic Press.

Calambokidis, J., Taylor, B. L., Carter, S. D., Steiger, G. H., Dawson, P. K. & Antrim, L. D. 1987. Distribution and haul-out behaviour of harbour seals in Glacier Bay, Alaska. *Canadian Journal of Zoology*, 65, 1391-1396.

Calupca, T. A., Fristrup, K. M. & Clark, C. W. 2000. A compact digital recording system for autonomous bioacoustic monitoring. *The Journal of the Acoustical Society of America*, 37, 13-22.

Charrier, I., Mathevon, N. & Jouventin, P. 2002. How does a fur seal mother recognize the voice of her pup? An experimental study of *Arctocephalus tropicalis*. *Journal of Experimental Biology*, 205, 613-622.

Clark, C. W. & Charif, R. A. 1998. Acoustic monitoring of large whales to the west of Britain and Ireland using bottom mounted hydrophone arrays, October 1996-September 1997. *Joint Nature Conservancy Report* No. 281.

Clark, C. W. 1995. Application of U.S. Navy underwater hydrophone arrays for scientific research on whales. *Scientific Report, International Whaling Commission*, 44, 210-213.

Clark, C. W., Charif, R., Mitchell, S. & Colby, J. 1996. Distribution and behaviour of bowhead bhale, *Balaena myticetus*, based on analysis of acoustic data collected during the 1993 spring migration off Point Barrow, Alaska. *Report of the International Whaling Commission*, 46, 541-552.

Clark, D. S., Flattery, J., Gisiner, R., Griffith, L., Schilling, J., Sledzinski, T. & Trueblood, R. 1994. Acoustic localization of whales in real time over large areas. *The Journal of the Acoustical Society of America*, 96, 3250.

Cleator, H., Stirling, I. & Smith, T. G. 1989. Underwater vocalizations of the bearded seal (*Erignathus barbatus*). *Canadian Journal of Zoology*, 67, 1900-1910.

Collins, K. T., Rogers, T. L., Terhune, J. M., McGreevy, P. D., Wheatley, K. E. & Harcourt, R. G. 2005. Individual variation f in-air female 'pup contact' calls in Weddell seals, *Leptonychotes weddellii*. *Behaviour*, 142, 167-189.

Collins, K. T., Terhune, J. M., Rogers, T. L., Wheatley, K. E. & Harcourt, R. G. 2006. Vocal individuality of in-air Weddell seal (*Leptonychotes weddellii*) pup 'primary' calls. *Marine Mammal Science*, 22, 933-951.

Collison, N. E. & Dosso, S. E. 2003. Acoustic tracking of a freely-drifting sonobuoy field: experimental results. *IEEE Journal of Oceanic Engineering*, 28, 554-561.

Comiso, J. C. & Gordon, A. L. 1987. Recurring polynyas over the Cosmonaut Sea and the Maud Rise. *Journal of Geophysical Research*, 92, 2819-2833.

Costa, D. P. 1993. The relationship between reproductive and foraging energetics and the evolution of the Pinnipedia. *Symposium of the Zoological Society of London*, 66, 293-314.

Dayton, P. K., Morbida, B. J. & Bacon, F. 1994. Polar marine communities. *American Zoologist*, 34, 90-99.

Erbe, C. 2000. Census of marine mammals. *SCOR Whaling Group 118*: New Technologies for observing marine life. November 9 - 11, 2000, Sidney, BC.

Eriksen, C. C., Osse, T. J., Light, R. D., Wen, T., Lehman, T. W., Sabin, P. L., Ballard, J. W. & Chiodi, A. M. 2001. Seaglider: a long-range autonomous underwater vehicle for oceanographic research. *IEEE Journal of Oceanic Engineering*, 26, 424-436.

Fay, F. H., Ray, G. C. & Kibal'chich, A. A. 1984. Time and location of mating and associated behavior of the Pacific walrus, *Odobenus rosmarus divergens* Illiger. *NOAA Technical Report NMFS*, 12, 89-99.

Figueroa, H. 2006. XBAT: Extensible Bioacoustic Tool. Retrieved (2008, January) from http://xbat.org.

Fletcher, S., Le Boeuf, B. J., Costa, D. P., Tyack, P. L. & Blackwell, S. B. 1996. Onboard acoustic recording from diving northern elephant seals. *The Journal of the Acoustical Society of America*, 100, 2531-2539.

Friedlaender, A. S., Johnston, D. W., Fink, S. L. & Lavigne, D. M. 2007. Variation in ice cover on the east coast of Canada, February-March, 1969-2006: implications for harp and hooded seals. International Fund for Animal Welfare Technical Report.

Fucile, P. D., Singer, R. C., Baumgartner, M. F. & K., B. 2006. A self contained recorder for acoustic observations from AUV's. MTS/IEEE Oceans '06. Boston, MA.

Gillespie, D. 2004. Detection and classification of right whale calls using an edge detector operating on a smoothed spectrogram. *Canadian Acoustics*, 32, 39-47.

Gisiner, R. & Schusterman, R. J. 1991. California sea lion pups play an active role in reunions with their mothers. *Animal Behaviour*, 41, 364-366.

Green, K. & Burton, H. R. 1988. Annual and diurnal dariations in the underwater vocalizations of Weddell seals. *Polar Biology*, 8, 161-164.

Greene, C. R., Jr, McLennan, M. W., Norman, R. G., McDonald, T. L., Jakubczak, R. S. & Richardson, W. J. 2004. Directional frequency and recording (DIFAR) sensors in seafloor recorders to locate calling bowhead whales during their fall migration. *The Journal of the Acoustical Society of America*, 116, 799-813.

Haller, M. A., Kovacs, K. M. & Hammill, M. O. 1996. Maternal behaviour and energy investment by grey seals breeding on land-fast ice. *Canadian Journal of Zoology*, 74, 1531-1541.

Hammill, M. O., Kovacs, K. M. & Lydersen, C. 1994. Local movements by nursing bearded seal (*Erignathus barbatus*) pups in Kongsfjorden, Svalbard. *Polar Biology*, 14, 569-570.

Hammill, M. O., Lydersen, C., Ryg, M. & Smith, T. G. 1991. Lactation in the ringed seal (*Phoca hispida*). *Canadian Journal of Fisheries and Aquatic Sciences*, 48, 2471-2476.

Hanggi, E. B. & Schusterman, R. J. 1994. Underwater acoustic displays and individual variation in male harbour seals, *Phoca vitulina*. *Animal Behaviour*, 48, 1275-1283.

Harcourt, R. G., Kingston, J. J., Cameron, M. F., Waas, J. R. & Hindell, M. A. 2007. Paternity analysis shows experience, not age, enhances mating success in an aquatically mating pinniped the Weddell seal (*Leptonychotes weddellii*). *Behavioural Ecology and Sociobiology*, 61, 643-652.

Harcourt, R. G., Kingston, J. J., Waas, J. R. & Hindell, M. A. in press. Foraging while breeding: alternative mating strategies by male Weddell seals? *Aquatic Conservation: Marine and Freshwater Ecosystems.*

Hayes, S. A., Kumar, A., Costa, D. P., Mellinger, D. K., Harvey, J. T., Southall, B. L. & Le Boeuf, B. J. 2004. Evaluating the function of the male harbour seal roar through playback experiments. *Animal Behaviour*, 67, 1133-1139.

Hayes, S. A., Mellinger, D. K., Croll, D. A., Costa, D. P. & Borsani, J. F. 2000. An inexpensive passive acoustic system for recording and localizing wild animal sounds. *The Journal of the Acoustical Society of America*, 107, 3552-3555.

Hindell, M. A., Harcourt, R. G., Waas, J. R. & Thompson, D. 2002. Fine-scale three-dimensional spatial use by diving, lactationg female Weddell seals, *Leptonychotes weddellii*. *Marine Ecology Progress Series*, 242, 275-284.

Insley, S. J. 1992. Mother-offspring separation and acoustic stereotypy: a comparison of call morphology in two species of pinnipeds. *Behaviour*, 120, 103-122.

Insley, S. J. 2001. Mother-offspring vocal recognition in northern fur seals is mutual but asymetrical. *Animal Behaviour*, 61, 129-137.

Insley, S. J., Phillips, A. V. & Charrier, I. 2003. A review of social recognition in pinnipeds. *Aquatic Mammals*, 29, 181-201.

Janik, V. M. 2000. Food-related bray calls in wild bottlenose dolphins (*Tursiops truncatus*). *Proceedings of the Royal Society London: Biological Sciences*, 267, 923-927.

Job, D. A., Boness, D. J. & Francis, J. M. 1995. Individual variation in nursing vocalizations of Hawaiian monk seal pups, *Monachus schauinsulandi* (Phocidae, Pinnipedia) and lack of maternal recognition. *Canadian Journal of Zoology*, 73, 975-983.

Johnson, M. P. & Tyack, P. L. 2003. A digital acoustic recording tag for measuring the response of wild marine mammals to sound. *IEEE Journal of Oceanic Engineering*, 28, 3-12.

Juang, B. H. & Rabiner, L. R. 1991. Hidden Markov Models for speech recognition. *Technometrics,* 33, 251-272.

Kastelein, R. A. 2002. Walrus. In: *Encyclopedia of Marine Mammals* (Ed. by Perrin, W. F., Wursig, B. & Thewissen, J. G. M.), pp. 1294-1300. London: Academic Press.

Kastelein, R. A., Postma, J. & Verboom, W. C. 1995. Airborne vocalizations of Pacific walrus pups (*Odobenus rosmarus divergens*). In: *Sensory systems of Aquatic Mammals* (Ed. by Kastelein, R. A., Thomas, J. A. & Nachtigall, P. E.), pp. 265-285. Woerden, The Netherlands: De Spil Publishers.

Kaufman, G. W., Siniff, D. B. & Reichle, R. 1975. Colony behaviour of Weddell seals, *Leptonychotes weddellii*, at Hutton Cliffs, Antarctica. *Rapports et Proces-Verbaux des Reunions, Conseil International pour L'Exploration scientifique de la Mer*, 169, 228-246.

Kelly, B. P. & Wartzok, D. 1996. Ringed seal diving behavior in the breeding season. *Canadian Journal of Zoology*, 74, 1547-1555.

Kibal'chich, A. A. & Lisitsina, T. Y. 1979. Some acoustical signals of calves of the Pacific walrus (in Russian). *Zoologicheskii zhurnal*, 58, 1247-1249.

Kindermann, L., Klinck, H. & Boebel, O. 2006. Automated detection of marine mammals in the vicinity of RV Polarstern. *Berichte zur Polar- und Meeresforschung*, 533, 68-76.

Klinck, H. 2008. Automated detection, localization and identification of leopard seals. Phd thesis. University of Trier, Germany.

Klinck, H., Kindermann, L. & Boebel, O. in press. Detection of leopard seal (Hydrurga leptonyx) vocalizations using the Envelope Spectrogram technique (tEST) in combination with a Hidden Markov Model. *Canadian Acoustics.*

Kooyman, G. L. 1981. Weddell seal: Consummate diver. New York: Cambridge University Press.

Kovacs, K. M. 1987. Maternal behaviour and early ontogeny of harp seals, *Phoca groenlandica*. *Animal Behaviour*, 35, 844-855.

Kovacs, K. M. 1990. Mating strategies in male hooded seals (*Cystophore cristata*). *Canadian Journal of Fisheries and Aquatic Sciences*, 68, 2499-2502.

Kovacs, K. M. 1995. Mother-pup reunions in harp seals, *Phoca groenlandica*: cues for the relocation of pups. *Canadian Journal of Zoology*, 73, 843-849.

Krafft, B. A., Lydersen, C., Kovacs, K. M., Gjertz, I. & Haug, T. 2000. Diving behaviour of lactating bearded seals (*Erignathus barbatus*) in the Svalbard area. *Canadian Journal of Zoology*, 78, 1408-1418.

Kunnasranta, W., Hyvarinen, H. & Sorjonen, J. 1996. Underwater vocalizations of Ladoga ringed seals (*Phoca hispida Ladogensis* Nordq) in summertime. *Marine Mammal Science*, 12, 611-618.

Laws, R. M. 1981. Biology of Antarctic seals. *Science Progress*, 67, 377-397.

Laws, R. M. 1984. Seals. In: *Antarctic Ecology* (Ed. by Laws, R. M.), pp. 621-715. London: Academic Press.

Le Boeuf, B. J. 1986. Sexual strategies of seals and walruses. *New Scientist*, 1491, 36-39.

Lopatka, M., Adam, O., Laplanche, C., Zarzycki, J. & Motsch, J. F. 2005. An attractive alternative for sperm whale click detection using the wavelet transform in comparison to the Fourier spectrogram. *Aquatic Mammals*, 31, 463-467.

Lopatka, M., Adam, O., Laplanche, C., Motsch, J.-F. & Zarzycki, J. 2006. Sperm whale click analysis using a recursive time-variant lattice filter. *Applied Acoustics*, 67, 1118.

Lydersen, C. & Hammill, M. O. 1993. Activity, milk intake and energy consumption in free-living ringed seal (*Phoca hispida*) pups. *Journal of Comparative Physiology B*, 163, 433-438.

Lydersen, C. & Kovacs, K. M. 1993. Diving behaviour of lactating harp seal, *Phoca groenlandica*, females from the Gulf of St Lawrence, Canada. *Animal Behaviour*, 46, 1213-1221.

Lydersen, C. & Kovacs, K. M. 1999. Behaviour and energetics of ice-breeding, North Atlantic phocid seals during the lactation period. *Marine Ecology Progress Series*, 187, 265-281.

Madsen, P. T., Payne, R., Kristiansen, N. U., Wahlberg, M., Kerr, I. & Mohl, B. 2002. Sperm whale sound production studied with ultrasound time/depth-recording tags. *Journal of Experimental Biology*, 205, 1899-1906.

Mathews, E. A. & Kelly, B. P. 1996. Extreme temporal variation in harbor seal (*Phoca vitulina Richardsi*) numbers in Glacier Bay, a glacial fjord in Southeast Alaska. *Marine Mammal Science*, 12, 483-489.

McCulloch, S. & Boness, D. J. 2000. Mother-pup vocal recognition in the grey seal (*Halichoerus grypus*) of Sable Island, Nova Scotia, Canada. *Journal of Zoology London*, 251, 449-455.

McCulloch, S., Pomeroy, P. P. & Slater, P. J. B. 1999. Individually distinctive pup vocalizations fail to prevent allo-suckling in grey seals. *Canadian Journal of Zoology*, 77, 716-723.

McDonald, M. A. & Fox, C. G. 1999. Passive acoustic methods applied to fin whale population density estimation. *The Journal of the Acoustical Society of America*, 105, 2643-2651.

Mellinger, D. K. 2001. Ishmael 1.0 User's guide. NOAA Technical Memorandum OAR PMEL-120, available from NOA/PMEL/OERD, 2115 SE OSU Drive, Newport, OR 97365-5258.

Mellinger, D. K. & Clark, C. W. 2000. Recognizing transient low-frequency whale sounds by spectrogram correlation. *The Journal of the Acoustical Society of America*, 107, 3518-3529.

Mellinger, D. K., Stafford, K. M., Moore, S. E., Dziak, R. P. & Matsumoto, H. 2007. An overview of fixed passive acoustic observation methods for cetaceans. *Oceanography*, 20, 36-45.

Merdsøy, B. R., Curtsinger, W. R. & Renouf, D. 1978. Preliminary underwater observations of the breeding behavior of the harp seal (*Pagophilus groenlandicus*). *Journal of Mammology*, 59, 181-185.

Miller, E. H. & Boness, D. J. 1983. Summer behavior of Atlantic walruses, *Odobenus rosmarus rosmarus* (L.), at Coats Island, N.W.T. (Canada). *Zeitschrift fuer Saeugetierkunde*, 48, 298-313.

Miller, E. H. & Murray, A. V. 1995. Structure, complexity and organization of vocalizations in harp seal (*Phoca groenlandica*) pups. In: *Sensory systems of Aquatic Mammals* (Ed. by

Kastelein, R. A., Thomas, J. A. & Nachtigall, P. E.), pp. 237-264. Woerden, The Netherlands: De Spil Publishers,

Miller, E. H. 1985. Airborne acoustic communication in Walrus, *Odobenus rosmarus*. *National Geographic Research*, 1, 124-145.

Møhl, B., Terhune, J. M. & Ronald, K. 1975. Underwater calls of the harp seal, *Pagophilus groenlandicus*. *Rapports et Proces-Verbaux des Reunions, Conseil International pour L'Exploration scientifique de la Mer*, 169, 533-543.

Moore, S. E., Stafford, K. M., Dalheim, M. E., Fox, C. G., Braham, H. W., Polovina, J. J. & Bain, D. E. 1998. Seasonal variation in fin whale call reception at five SOSUS sites in the North Pacific. *Marine Mammal Science*, 14, 617-626.

Moore, S. E., Stafford, K. M., Mellinger, D. K. & Hildebrand, J. A. 2006. Listening for large whales in the offshore waters of Alaska. *Bioscience*, 56, 49-55.

Newcomb, J., Fisher, R., Field, R., Turgut, A., Ioup, G., Rayborn, G., Kuczaj, S., Caruthers, J., Goodman, R. & Sidorovskaia, N. 2002. Near-bottom hydrophone measurements of ambient noise and sperm whale vocalizations in the northern Gulf of Mexico. *The Journal of the Acoustical Society of America*, 111, 2371.

Norris, T. F., McDonald, M. A. & Barlow, J. 1999. Acoustic detections of singing humpback whales (*Megaptera novaeangliae*) in the east North Pacific during their northbound migration. *The Journal of the Acoustical Society of America*, 106, 506-514.

Oetelaar, M. L., Terhune, J. M. & Burton, H. R. 2003. Can the sex of a Weddell seal (*Leptonychotes weddellii*) be identified by its surface call? *Aquatic Mammals*, 29, 261-267.

Oswald, J. N., Rankin, S. & Barlow, J. 2004. The effect of recording and analysis bandwidth on acoustic identification of delphinid species. *Journal of the Acoustical Society of America*, 116, 3178-3185.

Owens, W. B. 2006. Development and field testing of autonomous profiling floats across the Arctic ocean. *Eos Transactions, American Geophysical Union*, 87(36), Ocean Sciences Meeting Supplement. Honolulu, Hawaii. February 20-24, 2006. Abstract OS25N-12.

Perry, E. A. & Renouf, D. 1988. Further studies of the role of harbour seal (*Phoca vitulina*) pup vocalizations in preventing separation of mother-pup pairs. *Canadian Journal of Zoology*, 66, 934-938.

Perry, E. A. & Terhune, J. M. 1999. Variation of harp seal underwater vocalizations among three breeding locations. *Journal of Zoology London*, 249, 181-186.

Pettit, R., Harris, D., Wooding, B., Bailey, J., Jolly, J., Hobart, E., Chave, A., Duennebier, F., Butler, R., Bowen, A. & Yoerger, D. 2002. The Hawaii-2 Observatory. *IEEE Journal of Oceanic Engineering*, 27, 245-253.

Phillips, A. V. & Stirling, I. 2000. Vocal individuality in mother and pup South American fur seals, *Arctocephalus australis*. *Marine Mammal Science*, 16, 592-616.

Pierotti, R. & Pierotti, D. 1980. Effects of cold climate on the evolution of pinniped breeding systems. *Evolution,* 34, 494-507.

Rankin, S., Ljungblad, D., Clark, C. & Kato, H. 2005. Vocalisations of Antarctic blue whales, *Balaenoptera musculus intermedia*, recorded during the 2001/2002 and 2002/2003 IWC/SOWER circumpolar cruises, Area V, Antarctica. *Journal of Cetacean Research and Management*, 7, 13-20.

Renouf, D. 1984. The vocalization of the harbour seal pup and its role in the maintenance of contact with the mother. *Journal of Zoology London*, 202, 583-590.

Richardson, W. J., Greene, C. R., Jr, Malme, C. I. & Thomson, D. H. 1995. Marine mammals and noise. London: Academic Press.

Riedman, M. 1990. The pinnipeds. Seals, sea lions and walruses. Oxford: University of California Press.

Rogers, T. L. & Bryden, M. M. 1997. Density and haul-out behavior of leopard seals (*Hydrurga leptonyx*) in Prydz Bay, Antarctica. *Marine Mammal Science*, 13, 293-302.

Rogers, T. L. 2003. Factors influencing the acoustic behaviour of male phocid seals. *Aquatic Mammals*, 29, 247-260.

Rogers, T. L. 2007. Age-related differences in the acoustic characteristics of male leopard seals, *Hydrurga leptonyx*. *The Journal of the Acoustical Society of America*, 122, 596-605.

Rogers, T. L., Cato, D. H. & Bryden, M. M. 1996. Behavioral significance of underwater vocalizations of captive leopard seals, *Hydrurga letonyx*. *Marine Mammal Science*, 12, 414-427.

Rouget, P. A., Terhune, J. M. & Burton, H. R. 2007. Weddell seal underwater calling rates during the winter and spring near Mawson station, Antarctica. *Marine Mammal Science*, 23, 508-523.

Sato, K., Mitani, Y., Kusagaya, H. & Naito, Y. 2003. Synchronous shallow dives by Weddell seal mother-pup pairs during lactation. *Marine Mammal Science*, 19, 384-395.

Schusterman, R. J. 1970. Vocal displays underwater by the grey seal, harbour seal and the Steller seal lion. *Psychonomic Science*, 18, 303-305.

Seibert, A. 2007. The Ross seal and its underwater vocalizations. MSc thesis. University of Munich, Germany.

Sergeant, D. E. 1974. A rediscovered whelping population of hooded seals, *Cystophora cristata*, and its possible relationship to other populations. *Polarforschung*, 44, 1-7.

Serrano, A. & Miller, E. H. 2000. How vocal are harp seals (*Pagophilus groenlandicus*)? A captive study of seasonal and diel patterns. *Aquatic Mammals*, 26, 253-259.

Serrano, A. 2001. New underwater and aerial vocalizations of captive harp seals (*Pagophilus groenlandicus*). *Canadian Journal of Fisheries and Aquatic Sciences*, 79, 75-81.

Sherman, J., Davis, R. E., Owens, W. B. & Valdes, J. 2001. The autonomous underwater glider 'Spray'. *IEEE Journal of Oceanic Engineering*, 26, 437-446.

Simard, Y., Park, C., Bahoura, M., Rouat, J., Sirois, M., Mouy, X., Seebaruth, D., Roy, N. & Lepage, R. 2006. Development and experimentation of a satellite buoy network for real-time acoustic localisation of whales in the St Lawrence MTS/IEEE Oceans '06. Boston, MA.

Siniff, D. B., Stirling, I., Bengston, J. L. & Reichle, R. A. 1979. Social and reproductive behavior of crabeater seals during the austral spring. *Canadian Journal of Zoology*, 57, 2243-2255.

Sirovic, A., Hildebrand, J. A., Wiggins, S. M., Moore, S. E., McDonald, M. A. & Thiele, D. 2003. Seasonality of blue and fin whale calls west of the Antarctic Peninsula. *Oceans 2003 Proceedings*, 2, 818.

Sjare, B. & Stirling, I. 1996. The breeding behavior of Atlantic walruses, *Odobenus rosmarus rosmarus*, in the Canadian High Arctic. *Canadian Journal of Zoology*, 74, 897-911.

Smith, T. G. & Hammill, M. O. 1981. Ecology of the ringed seal, *Phoca hispida*, in its fast-ice breeding habitat. *Canadian Journal of Zoology*, 59, 966-981.

Spikes, C. H. & Clark, C. W. 1996. Whales 95 - Revolutionizing marine mammal monitoring technology. *Sea Technology*, 37, 49-56.

Stacey, R. M. 2006. Airborne and underwater vocalizations of the Antarctic Ross seal (Ommatophoca rossii). MSc thesis. Western Illinois University, Macomb, Illinois

Stafford, K. M. 1995. Characterization of blue whale calls from the northeast Pacific and development of a matched filter to locate blue whales on U.S. Navy SOSUS arrays. MSc thesis Oregon State University, Corvallis, Oregon.

Stafford, K. M., Fox, C. G. & Clark, D. S. 1998. Long-range acoustic detection and localization of blue whale calls in the northeast Pacific using military hydrophone arrays. *The Journal of the Acoustical Society of America,* 104, 3616-3625.

Stafford, K. M., Nieukirk, S. L. & Fox, C. G. 1999. Low-frequency whale sounds recorded on hydrophones moored in the eastern tropical Pacific. *The Journal of the Acoustical Society of America*, 106, 3687-3698.

Stirling, I. & Siniff, D. B. 1979. Underwater vocalizations of leopard seals (*Hydrurga leptonyx*) and crabeater seals (*Lobodon carcinophagus*) near the South Shetland Islands, Antarctica. *Canadian Journal of Zoology*, 57, 1244-1248.

Stirling, I. & Thomas, J. A. 2003. Relations between underwater vocalisations and mating systems in phocid seals. *Aquatic Mammals*, 29, 227-246.

Stirling, I. 1973. Vocalization of the ringed seal (*Phoca hispida*). *Journal of the Fisheries Research Board of Canada*, 30, 1592-1594.

Stirling, I. 1975. Factors affecting the evolution of social behaviour in the pinnipedia. *Rapports et Proces-Verbaux des Reunions, Conseil International pour L'Exploration scientifique de la Mer*, 169, 205-212.

Stirling, I., Calvert, W. & Cleator, H. 1983. Underwater vocalizations as a tool for studying the distribution and relative abundance of wintering pinnipeds in the high Arctic. *Arctic*, 36, 262-274.

Stirling, I., Calvert, W. & Spencer, C. 1987. Evidence of stereotyped underwater vocalizations of male Atlantic walruses, *Odobenus rosmarus rosmarus*. *Canadian Journal of Zoology*, 65, 2311-2321.

Stonehouse, B. 1989. Polar environments. In: *Polar Ecology*. New York: Chapman and Hall.

Streveler, G. P. 1979. Distribution, population ecology, and impact susceptibility of the harbor seal in Glacier Bay, Alaska. Final report to the U.S. National Park Service. Juneau, AK.

Tedman, R. A. & Bryden, M. M. 1979. Cow-pup behaviour of the Weddell seal, *Leptonychotes weddellii* (Pinnipedia), in McMurdo Sound, Antarctica. *Australian Wildlife Research*, 6, 19-37.

Terhune, J.M. 1994. Geographical variation of harp seal underwater vocalizations. *Canadian Journal of Zoology*, 72, 892-897.

Terhune, J. M. & Dell'Apa, A. 2006. Stereotyped calling patterns of a male Weddell seal (*Leptonychotes weddellii*). *Aquatic Mammals*, 32, 175-181.

Terhune, J. M. & Ronald, K. 1986. Distant and near-range functions of harp seal underwater calls. *Canadian Journal of Fisheries and Aquatic Sciences*, 64, 1065-1070.

Terhune, J. M., Stewart, R. E. A. & Ronald, K. 1979. Influence of vessel noises on underwater vocal activity of harp seals. *Canadian Journal of Zoology*, 57, 1337-1338.

Thomas, J. A. & DeMaster, D. P. 1982. An acoustic technique for determinig diurnal activities in leopard (*Hydrurga leptonyx*) and crabeater (*Lobodon carcinophagus*) seal. *Canadian Journal of Zoology*, 60, 2028-2031.

Thomas, J. A. & Kuechle, V. B. 1982. Quantitative analysis of Weddell seal (*Leptonychotes weddelli*) underwater vocalizations at McMurdo Sound, Antarctica. *The Journal of the Acoustical Society of America*, 72, 1730-1738.

Thomas, J. A. & Stirling, I. 1983. Geographic variation in the underwater vocalizations of Weddel seals (*Leptonychotes weddelli*) from Palmer Peninsula and McMurdo Sound, Antarctica. *Canadian Journal of Zoology*, 61.

Thomas, J. A., Zinnel, K. C. & Ferm, L. M. 1983. Analysis of Weddell seal (*Leptonychotes weddelli*) vocalizations using underwater playbacks. *Canadian Journal of Zoology*, 61, 1448-1456.

Tregenza, N. J. C. 1999. A new tool for cetacean research - a fully submersible click detector. In: European Research on Cetaceans 13. (Ed. by P.G.H. Evans, J. Cruz and J.A. Raga). Proceedings of the 13th Annual ECS Conference, Valencia, Spain, 5-8 April 1999.

Trillmich, F. 1981. Mutual mother-pup recognition in Galapagos fur seals and sea lions: cues used and functional significance. *Behaviour*, 78, 21-42.

Trillmich, F. 1996. Parental investment in pinnipeds. *Advances in the Study of Behaviour*, 25, 533-577.

Tynan, C. T. & DeMaster, D. P. 1997. Observations and predictions of Arctic climate change: potential effects on marine mammals. *Arctic*, 50, 308-322.

Van Opzeeland, I. C. & Van Parijs, S. M. 2004. Individuality in harp seal (Phoca groenlandica) pup vocalizations. Animal Behaviour, 68, 1115-1123.

Van Opzeeland, I.C., Corkeron, P.J., Risch, D., Stenson, G.B. & Van Parijs. In prep. Variation in harp seal, Phoca groenlandica, pup vocalisations and behaviour between whelping patches in the Greenland Sea and Newfoundland, Canada.

Van Parijs, S. M. 2003. Aquatic mating in pinnipeds: a review. *Aquatic Mammals*, 29, 214-226.

Van Parijs, S. M., Hastie, G. D. & Thompson, P. M. 1999. Geographical variation in temporal and spatial vocalization patterns of male harbour seals in the mating season. *Animal Behaviour*, 58, 1231-1239.

Van Parijs, S. M., Hastie, G. D. & Thompson, P. M. 2000. Individual and geographical variation in display behaviour of male harbour seals in Scotland. *Animal Behaviour*, 59, 559-568.

Van Parijs, S. M., Kovacs, K.M. & Lydersen, C. 2001. Spatial and temporal distribution of vocalising male bearded seals-implications for male mating strategies. *Behaviour*, 138, 905-922.

Van Parijs, S. M., Lydersen, C. & Kovacs, K. M. 2004. Effects of ice cover on the behavioural patterns of aquatic mating male bearded seals. *Animal Behaviour*, 68, 89-96.

Van Parijs, S. M., Lydersen, C., Kovacs, K.M. 2003. Vocalizations and movements suggest alternative mating tactics in male bearded seals. *Animal Behaviour*, 65, 273-283.

Van Parijs, S. M., Thompson, P. M., Tollit, D. J. & Mackay, A. 1997. Distribution and activity of male harbour seals during the mating season. *Animal Behaviour*, 54, 35-43.

Van Parijs, S.M., Southall, B. & Workshop Co-chairs. 2007. Report of the 2006 NOAA National Passive Acoustics Workshop: Developing a Strategic Program Plan fro NOAA's Passive Acoustics Ocean Observing System (PAOOS), Woods Hole, Massachusetts, 11-

13 April 2006. U.S. Dep. Commerce, NOAA Technical Memorandum. NMFS-F/SPO-81, 64p.

Watkins, W. A. & Ray, G. C. 1977. Underwater sounds from ribbon seal, *Phoca (histriophoca) fasciata*. *Fisheries Bulletin*, 75, 450-453.

Watkins, W. A. & Ray, G. C. 1985. In-air and underwater sounds of the Ross seal, *Ommatophoca rossi*. *The Journal of the Acoustical Society of America*, 77, 1598-1600.

Watkins, W. A. & Schevill, W. E. 1979. Distinctive characteristics of underwater calls of the harp seal, *Phoca groenlandica*, during the breeding season. *The Journal of the Acoustical Society of America,* 66, 983-988.

Wiggins, S. M. 2003. Autonomous acoustic recording packages (ARPs) for long-term monitoring of whale sounds. *Marine Technology Society Journal*, 37, 13-22.

Wilson, S. 1974. Juvenile play of the common seal, *Phoca vitulina vitulina*, with comparative notes on the grey seal, *Halichoerus grypus*. *Behaviour*, 48, 37-60.

In: Animal Behavior: New Research
Editors: E. A. Weber, L. H. Krause

ISBN: 978-1-60456-782-3
© 2008 Nova Science Publishers, Inc.

Chapter 7

FOOD HOARDING IN THE NEW ZEALAND ROBIN: A REVIEW AND SYNTHESIS

Ignatius J. Menzies and K. C. Burns[*]
School of Biological Sciences, Victoria University of Wellington.
P.O. Box 600, Wellington, New Zealand

ABSTRACT

Experiments on avian food hoarding have played a prominent role in our understanding of animal behaviour. Important insights into cooperative breeding, foraging theory, social learning, spatial memory and territoriality have been generated by studies investigating how birds create, protect and retrieve caches. However, our understanding of avian food hoarding is limited to just a few groups of birds, namely acorn woodpeckers, shrikes, parids (tits and chickadees) and corvids (crows, jays and ravens). Given the breadth of information gained from a small number of model organisms, studies on different food hoarding birds may generate valuable new insights into animal behaviour. The New Zealand robin's food hoarding behaviour is unusual and poorly understood. Robins hunt some of the world's largest invertebrate prey, which they cannot consume whole. Large prey are instead dismembered and stored piecemeal in tree canopies. Like many other birds endemic to isolated islands, robins lack pronounced anti-predatory behaviours and are fearless of humans. Wild birds readily cache prey offered to them by hand and recent field experiments have shed light on some aspects of their food hoarding behaviour. Here, we synthesize current understanding in a review of food hoarding by the New Zealand robin. We begin with a brief review of robin life history, emphasizing the importance of food hoarding. Next we discuss the results of recent field experiments investigating the ecological and social factors regulating when, where and how robins hoard food. Lastly we discuss the uniqueness of New Zealand robins by emphasizing differences with other food-hoarding systems and how these differences may have arisen from New Zealand's unique ecological and evolutionary history.

---

[*] Phone: 64-4-463-5339; *Fax:* 64-4-463 5331; Email: kevin.burns@vuw.ac.nz

1. LIFE HISTORY OF THE NEW ZEALAND ROBIN

The New Zealand robin (*Petroica australis*) is a medium-sized, insectivorous passerine in the family Petroicidae [1]. The family Petroicidae has historically been incorrectly placed with the Eurasian robins (Old-World flycatchers, Muscicapidae). However, paleontological, phylogenetic and molecular data suggest Petroicidae are part of the Australo-Papuan radiation of oscines (parvorder Corvida *sensu* Sibley & Ahlquist [2]) and thought to be basal to the Afro-Eurasian songbirds (parvorder Passerida) [3, 4].

The family comprises approximately 44 species, most of which occur in temperate Australia. The New Zealand tomtit (*Petroica macrocephala*), the black robin (*Petroica traversi*) and the New Zealand robin are endemic to New Zealand [5]. The New Zealand robin has recently been split into two species; the North Island robin (*ssp. longipies*) and the South Island robin (*spp. australis*) [6-8]. However, for the purposes of this review we will consider them jointly.

New Zealand robins weigh about 35g and are around 18cm in length. Males and females are sexually dimorphic. They have a bold upright stance and long thin legs (Figure 1). The plumage is mostly grey-black with a whitish breast, varying in prominence with sex and subspecies [8]. Males have darker plumage on their back and upper breast than females [9]. Immature birds are similar in appearance to females and males show delayed plumage maturation, acquiring characteristic male plumage after their first breeding season, 12 to 16 months after fledging [9, 10]. Robins show no seasonal variation in plumage, other than during the moult which occurs around late summer.

Figure 1. A New Zealand robin (*Petroica australis*) 'following' human observers.

New Zealand robins usually form life-long pair bonds, although examples of mate switching and polyandry are sometimes observed [11]. Both sexes contribute to care of the young. Females are fed several times an hour by males while building the nest [12]. Two to four eggs are laid and incubated for around 18 days, during which the female is called off the nest and fed two to three times per hour [7]. However, nuptial feeding by the male mostly stops after the hatching of offspring [12, 13]. Both sexes feed the young [7]. However, when females begin re-nesting, the male takes responsibility for fledgling provisioning. Fledglings begin foraging approximately 2 weeks after leaving the nest [7], and are fed for 25 to 50 days before being encouraged to leave the territory [12].

New Zealand robins are highly territorial [14], especially in the breeding season. The male patrols his territory, singing loudly and aggressively confronting encroaching birds, raising crown feathers and sometimes flashing a white frontal spot in the centre of their forehead [7, 15-17]. Encounters with other robins or other species (fantail, silvereye and tomtit [18]) often result in violent pursuits and assaults [15, 16]. Male birds are socially dominant to females [15, 19] and outside the breeding season they can be aggressive towards females [20].

New Zealand robins eat mostly invertebrates and some berries, although they do not store berries [14]. Their diet includes some of the world's largest invertebrates and robins frequently catch prey that is longer and/or heavier than themselves [21, 22], such as giant earthworms (Lumbricidae [23]), stick insects (Phasmatodea [24]) and flightless giant grasshoppers called weta (Orthoptera: Anostostomatidae [25]). Once captured, large prey are typically dismembered and the pieces are consumed or stored in branch-truck axils in the forest canopy [14].

Robins spend 90% of their foraging time on or within 2m of the ground [22]. They often use a low branch, boulder or trunk as a vantage point from which to scan the ground, before flying down to catch their prey [22]. They also actively search for prey by turning-over leaves on the forest floor. Foot-trembling movements are used to stimulate movement from hidden prey [17, 22, 26]. Robins also glean prey from vegetation, take insects while hovering, and rarely catch flying prey on the wing [22].

New Zealand is one of the world's most isolated landmasses, which separated from Gondwana in the late Cretaceous (90 – 65 MYA) [27]. As it drifted away it took a subset of the existing Gondwanan fauna with it [28], a fauna dominated by invertebrate, amphibian, reptile and avian lineages [21, 29]. This separation occurred prior to the major mammal radiation of the Oligocene (37 – 24 MYA) [28]. The result is that, with the exception of Mystacinid bats, New Zealand flora and fauna evolved without the presence of any ground-dwelling mammals [28].

New Zealand's isolation, coupled with its lack of terrestrial mammals, has had important consequences for the evolution of its fauna. Many New Zealand animals adapted to niches that elsewhere were claimed by mammals. For example, weta have taken on the distinctive ecological niche of small mammals and are the only insects known to consume fleshy fruits and disperse the seeds thereafter, in a manner similar to small mammals elsewhere [30]. Many avian, invertebrate and reptilian lineages show convergent increases in body size. Flightlessness is also common; approximately 5 – 35% of avian species were flightless at the time of the first human colonisation of New Zealand [31].

Many species are also extremely 'naïve' (*sensu* Carlquist [32]) and show no innate recognition of alien, mammal predators such as stoats, cats and rats [28]. This attribute has led to severe declines in most bird species following human arrival [33]. Exotic mammals have had a profoundly negative impact on robins (see also Diamond & Veitch [34]). In some parts of New Zealand, robin populations are confined to offshore islands, which either escaped invasion by predatory mammals or have had invasive mammals removed via conservation efforts. However, North and South Island robins are still widespread through some areas of native and exotic forest in the central North Island, and both northern and southern portions of the South Island [35]. The species is currently listed in the IUCN Red List of Threatened Species [36].

2. FOOD HOARDING BEHAVIOUR

Faced with a temporary over-abundance in food, when New Zealand robins capture large prey they typically hoard what is not immediately consumed. Smaller prey items are usually consumed, except for when large numbers of small prey are caught over a short period of time. Under these circumstances, small prey are cached whole [14].

Robins typically prepare prey before transporting them to storage sites [22]. Large invertebrates are killed and dismembered by stabbing and pinching movements of the beak or are repeatedly and vigorously slammed against a hard object. Slimy prey such as slugs, snails and large earthworms are thoroughly rubbed on a log or over the ground presumably to remove slime that may foul the bird's plumage [22]. Smaller prey that are cached whole are also wounded, but are often still alive, which may delay spoilage. Usually a small portion of large prey is consumed before transporting the rest.

Most prey items are transported less than 10m from the place of preparation [14]. Items are cached in the forest canopy, usually in branch-trunk axils, stump ends, branch holes or on tree fern fronds (Figure 2). The mean height of cache sites (2.8m) is significantly higher than the mean height at which robins forage from vegetation (1.5m) (see also Powlesland [14]). Both sexes cache food. However, the threat of male pilferage often causes females to move out of sight of the male before caching [8]. Powlesland [14] observed that when robins foraged outside their territories in autumn, surplus food was cached very near to where it was found and not brought back within territorial boundaries.

Cache site selection often occurs hastily and food is usually stored in the first location visited [14]. In Powlesland's 1980 study the mean time taken to store an item was 24 seconds. Prey items are typically not wedged in place nor are they held in place by saliva (c.f. [37]) although many cached items suffer from puncture wounds and/or burst abdominal cavities which may adhere them to the cache site when coagulated [14]. Unlike most other food hoarding systems [38], no attempt is made to conceal or camouflage cache sites with bark or lichen [13, 14], and cached items are often visible from above [14]. Steer and Van Horik [39] observed bellbirds (*Anthornis melanura*) and stichbirds (*Notiomystis cincta*) stealing from robin caches.

Figure 2. (A) Typical forest trail in the Kaori Wildlife Sanctuary along which trials are conducted. (B) A robin during an experimental trial. The bird is holding several meal worms in its bill and is posed to pick-up another from the forest floor. (C) A typical cache site, filled with mealworm prey.

Pre-human New Zealand was home to a diverse array of large and aggressive ground-dwelling omnivores. Large, flightless predators and scavengers such as large flightless rails (e.g. weka, *Gallirallus australis*), reptiles (e.g. tuatara, *Sphenodon puncatus*) and the extinct adzebill (*Aptornis otidiformis*) would have made the forest floor a hazardous place for a small bird to spend time handling prey. It is likely that quickly transporting prey into tree canopies evolved as a behavioural strategy to reduce cache pilferage by larger animals.

Food storage in robins is a short term process and caches are usually retrieved within 1 to 3 days [14]. Invertebrate prey decompose relatively quickly, even in winter, and prey that remain in caches after three days may often be rejected due to spoilage [14]. The short cache duration may be related to this fast rate of decay. Robins store and retrieve food throughout the day but rates of prey consumption and retrieval increase later in the day [14]. Robins, like other small birds, have a high metabolic rate and towards the end of the day must consume enough food to meet both their immediate physiological requirements and sustain them through long, cool winter nights [40]. Powlesland [14] proposed that food stored during the day serves as a reliable, readily accessible source of energy that can be retrieved in the evening by robins to meet their energetic needs. He also noted that some caches remain available for retrieval early the next morning and may be used to provide energy at a time when invertebrates were likely to be least active and therefore difficult to detect. Of 40 caches monitored by Powlesland in 1980, 58% disappeared within the day they were stored, but the cause was not determined. Interspecific cache pilferage by birds and insects has been observed [39, 41]. Subsequent experiments have found high levels of re-caching and cache theft in robins [42] and are discussed below.

3. RECENT FOOD HOARDING EXPERIMENTS

i) Naiveté and Field Methods

The unusual life history attributes of New Zealand robins generates a unique opportunity for field experiments on the food hoarding behaviour of wild birds. Robins will fearlessly approach humans. They are extremely inquisitive and often follow humans walking through the bush. They seem to be particularly attracted to any disturbance to the leaf litter caused by hiking [7], leading to speculation that they may have benefited from disturbances created by large flightless ratites called moa, which were common prior to human arrival but are now extinct [43]. Robins readily inspect human activities and can be attracted to an experimenter by noises such as clapping, whistling or banging a tree with a stick [15]. As a result, activities that are difficult to document in most songbirds (e.g. mating and nuptial feeding) are readily observed in wild New Zealand robins [14, 22]. They will also consume and cache food offered to them by hand, providing a unique method of conducting field experiments on wild, untrained birds.

The experimental study of robin food hoarding behaviour is greatly facilitated by recent conservation efforts and in particular, the Karori Wildlife Sanctuary (KWS). KWS is an ambitious attempt to restore the native avifauna on New Zealand's larger islands, where total eradication of introduced mammals is infeasible. KWS is a 'mainland island' of regenerating bush within which introduced mammals have been removed, and reinvasion is prevented by a special mesh fence. In 2001, a population of robins was established by translocating birds from a nearby offshore island (Kapiti Island [44]). The species has since thrived in KWS, breeding prolifically and the population is now comprised of 100's of birds. Much of the population is colour-banded, allowing identification of individuals. KWS is easily accessible, located less than 2km from Victoria University of Wellington [45].

The first study dedicated to the food hoarding behaviour of New Zealand robins was published in 1980 by Ralph Powlesland. In this seminal paper he described two years of observations of food hoarding by wild South Island robins. The first experimental study was conducted 25 years later. Alexander et al. [13] offered food to robins and compared whether the behaviour of free-ranging birds was similar to that of hand-fed birds. Their protocol was simple. Experimenters traversed a series of trails in KWS (Figure 2.), and once a robin (or a pair of robins) was encountered, two grams of mealworms were placed on the ground in front of birds and their subsequent behaviour was recorded. Results from Alexander et al.'s [13] experiments showed the behaviour of hand-fed birds in KWS and Powlesland's and wild population on the South Island was similar (Figure 3). Therefore, there is no indication to date that the behaviour of hand-fed birds in KWS differs markedly from wild birds. All subsequent field experiments have utilised Alexander et al.'s [13] methodology.

ii) Annual Cycles in Food Hoarding Behaviour

Most organisms, especially those that live at higher latitudes, experience seasonal changes in resource availability and coordinate their behaviour according to these cycles [46]. For example, many animals coordinate sexual reproduction to coincide with periods when

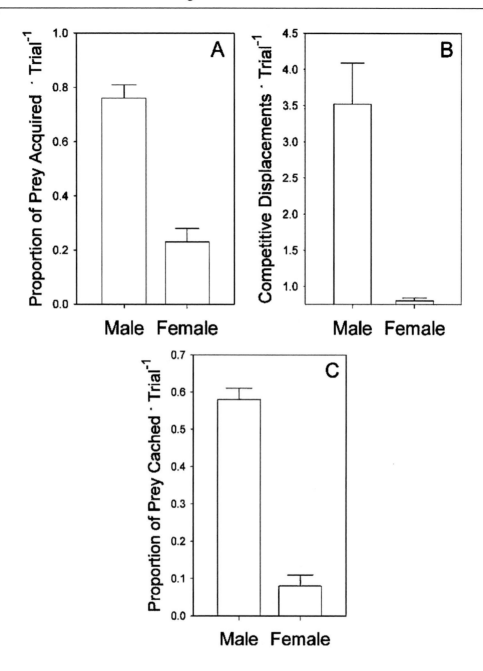

Figure 3. Sexual differences in food acquisition rates (A), aggressive encounters between mated pairs (B) and food hoarding rates (C) in winter. Redrawn from [12] with permission.

resources are most abundant [47]. Many food hoarding animals store food according to seasonal cycles [38]. Peak food hoarding usually occurs when food resources are most abundant. For example, many species of tits, titmice and chickadees (Paridae) store thousands of seeds on winter territories [48]. Studies on food hoarding have been conducted almost exclusively during seasonal peaks in food storage and retrieval, and as a result food hoarding behaviour outside of this peak is incompletely understood for most temperate songbirds and deserves investigation [49].

Using Alexander *et al*.'s [13] methodology, Steer and Burns [50] documented annual cycles of food use by robins for two years. Territorial pairs were given two grams of mealworms at monthly intervals. Subsequent behavioural observations were used to document seasonal changes in intersexual aggression, food acquisition, food consumption and food hoarding. They found that conflict between sexes occurs year-round. Males aggressively monopolise food sources and consistently acquire more mealworms than females [50]. Females occasionally gain access to food sources directly when males are away caching food, feeding juveniles or defending territorial boundaries. Upon returning to food sources, males always displace females in aggressive encounters that may sometimes end in injury. Aggression is less pronounced in the breeding season (September to February) [50]. However, males never relinquish control of artificial food sources, and presumably natural food sources as well. Although males physically monopolize food sources year-round, females gain access to artificial prey in a variety of ways that vary seasonally.

When males and females cooperate to raise young during the austral summer, males feed large amounts of prey to females and, as a consequence, cache far less [50]. This is sometimes accompanied by a food acquisition display, which is discussed further in section 4. Both sexes contribute to the care of fledglings, and males feed the young at approximately the same rate as females. The provisioning of young ends in autumn as does mate provisioning by males. Females hoard more food than males in the summer breeding season, while males hoard more food in the winter, non-breeding season [50]. Greater rates of food hoarding by females in summer are subsidized by nuptial feeding by males. Females often hoard food given to them by males close to nest sites, making cached food easy to retrieve during the nesting period. Later in the breeding season, females also use caches to feed fledglings.

The situation changes markedly in winter. As temperatures drop, food availability dwindles and daylight foraging times decline, males become more aggressive. Males cease feeding young and their mate. They instead begin caching all excess food [50]. Females cache far less in winter and instead consume nearly all of the limited amount of food that they acquire [50]. However, this does not mean that they ultimately obtain less prey from artificial food sources than males in winter. Females adopt an alternative food acquisition strategy that involves stealing food from their mate's caches.

iii) Intersexual Competition

Birds can store food in two ways. First, food can be consumed, converted to fat and stored internally. Alternatively, it can be stored externally as caches [51]. Both storage mechanisms have costs and benefits (see [52, 53]). Internally stored energy cannot be lost to pilferage [54]. However, heavier birds are slower and less agile thus carry a higher risk of predation [55]. They also require more energy to conduct other activities such as territory defence and foraging [56-58]. Birds can maintain lower body mass by caching food externally; and in a recent review, Brodin [59] suggested that the mere presence of cached food allows small birds to maintain lower fat stores, regardless of whether the food is retrieved or not. However, both storing and retrieving caches requires time and energy.

Additionally, food stored in caches can spoil [54], be lost to theft [60, 61], and their locations can be forgotten [62].

Decisions about food storage are also thought to be influenced by an animal's social environment. For many birds, accessibility to food and the security of caches is affected by the company of other birds [63]. Socially dominant birds are able to physically displace competitively subordinate birds from food sources [64, 65]. When in the presence of dominant birds, subordinate birds have less certain access to food sources. Therefore, subordinate birds are predicted to store greater quantities of food to offset losses to dominants [52, 62, 63]. Accordingly, it may be more profitable for dominant birds to store less food [63] and spend more time pilfering food from subordinate birds, rather than storing their own food. Support for these predictions is mixed [66].

Burns and Steer [19] investigated how intersexual competition influences New Zealand robin food hoarding behaviour. They presented mealworms to lone and paired robins to evaluate whether one sex was socially dominant, whether the subordinate bird cached more frequently than the dominant, and if birds cached more when in the presence of a potential competitor. These results showed that males dominate experimental food sources, aggressively defending them against females, and consequently they acquire over twice as many mealworms. Dominant males made twice as many caches as females and both sexes cached less in the presence of their mate. Therefore, New Zealand robin caching behaviour is contrary to predictions of food hoarding theory [52, 62, 63]. However, Brodin et al.'s [63] predictions were based upon seed-caching bird species that form large flocks in winter. Burns and Steer's [19] results suggest that these predictions may not be applicable to New Zealand robin, an insectivore that resides in pairs upon permanent territories.

Although males cached more than females and solitary birds cached more than paired birds, males and females cached at different rates in different social situations [19]. Females showed a dramatic drop in caching rates in the presence of males, while a less pronounced drop was observed in males [19]. This suggests that males are less susceptible to cache pilferage than females. Males often stay close to their caches and attempt to defend them against females. Conversely, females appear to be unable to protect their caches from males. The greater reduction in caching rates of paired females might therefore be an adaptive response to increased rates of pilfering by males. Females may cache fewer worms to avoid external storage costs, particularly cache pilferage by males. However, subsequent work would argue against the subordinate status of females during cache retrieval.

iv) Cache Spacing Patterns and Reciprocal Cache Theft

The "selfish-hoarding hypothesis" predicts that food hoarding is adaptive only when the animals that create caches have a higher probability of retrieving them [67, 68] (but see [69]). Although unambiguous tests of this hypothesis are rare [70-72], in a recent review Vander Wall & Jenkins [61] found that cache theft is common, and that many food-hoarding animals rarely recover the majority of their caches.

Many animals that store food externally display cache protection behaviours in order to reduce cache theft [73]. When in the presence of potential thieves, some animals modify their behaviour in ways that make pilfering more difficult [74, 75] or avoid caching altogether [62,

76-78]. Others re-cache food in new locations [79, 80]. Altering the spatial location of cache sites to make caches more difficult to locate or access might also reduce cache theft [81, 82].

Socially subordinate birds, which are unable to defend their caches, often hoard small amounts of food in large numbers of highly dispersed sites. This 'scatter hoarding' makes random location by thieves less likely and reduces losses when caches are found [64, 81, 83-85]. Conversely, dominant birds may benefit from 'larder hoarding', namely storing large amounts of food in a smaller number of sites that are located in close proximity, which can be more easily defended [62, 86-89]. However, it is unclear whether cache spacing patterns actually facilitate cache retrieval and discourage cache theft.

Several models have been proposed to explain the existence of cache protection strategies despite the high levels of cache theft. Vander Wall and Jenkins [61] suggested that because many pilferers also create caches, 'reciprocal' cache pilferage could explain the evolution of caching despite high levels of pilferage. Reciprocal cache use may also be explained by kin selection. For example, acorn woodpeckers, *Melanerpes formicivorus* [90] and beavers, *Castor canadensis* [91] form communal larders that are jointly constructed, used and defended by family groups. Cache pilferage might also be tolerated as an indirect form of mate provisioning (cf. [92]). A monogamous, territorial food hoarder may tolerate cache theft if its caches are stolen by its mate [93] as the stolen food might enhance an animal's inclusive fitness by indirectly provisioning the hoarder's offspring.

Van Horik and Burns [94] evaluated relationships between cache spacing patterns and cache retrieval in New Zealand robins. Mealworms were fed to male and female robins either when they were alone or when they were accompanied by their partner. Cache spacing patterns, cache retrieval rates and rates of cache theft were quantified and compared between sexes and social context to test three hypotheses: 1) socially subordinate females create more widely dispersed caches than socially dominant males, 2) cache spacing patterns are accentuated by the presence of a bird's mate relative to when foraging alone, and 3) spatial patterns in cache dispersal facilitate cache retrieval and reduce cache theft.

Spatial patterns in cache location were consistent with the hypothesis that dominant males aggregate their caches spatially, while subordinate females disperse their caches more widely. However, sexual differences in cache location were context dependent. When in the presence of their mates, females stored food at greater distances from food sources [94]. But the number of cache cites created by females was higher than that of males, regardless of social condition [94].

Similar results for cache spacing patterns have been found in other food-hoarding birds, such as white-breasted nuthatches (*Sitta carolinensis*) and several species of both parids and corvids [79, 82, 95-97], which are usually interpreted as behavioural adaptations to reduce cache theft [73, 75, 79, 88, 98-101]. By aggregating cache sites, defence by dominant animals may become easier as travel time between food stores is reduced [102-106]. Conversely, scatter hoarding may reduce the likelihood of discovery by pilferers using area-restricted searches [62, 82, 107]. A similar explanation could apply to New Zealand robins. Males aggregated cache sites and aggressively defended them, while females scattered undefended cache sites more widely. Nevertheless, cache retrieval data provides no support for this interpretation.

Cache theft is reciprocal in New Zealand robins (sensu Vander Wall & Jenkins [61]). Both sexes steal from their mate at least as frequently as they retrieved their own caches. This result is inconsistent with the predictions of the selfish hoarding theory, which argues that hoarding is adaptive only when the hoarder is more likely to retrieve its caches than other animals [67]. Although both males and females stole from their mate's caches, females were more successful pilferers and stole more food from male-made caches than vice versa [94]. This is surprising given that males aggregate their cache sites and frequently attacked females in cache defence.

Why do robins show pronounced cache spacing patterns if they fail to provide a retrieval advantage? There are several possible explanations. The inability of males to successfully defend their caches may result from the behavioural requirements of territory maintenance. Females typically gain access to male-made caches when males depart to conduct other activities, such as defending territory boundaries against neighbouring males. Females take no part in territory defence, which may allow them to spend more time pilfering male-made caches. Alternatively, high levels of cache theft may be tolerated as an indirect form of mate provisioning. Although food pilfered by females in early winter is unlikely to directly enhance reproductive success, pilfered food may prevent mate switching, or enhance female survivorship, which would be advantageous to males if pairs with past breeding experience have a better chance of successfully fledging young [108]. To better understand the mechanisms underpinning reciprocal cache theft, additional experiments are needed to evaluate relationships between male aggression, cache spacing patterns and cache theft as winter turns to spring.

Retrieving and then relocating previously cached food is a common cache protection strategy. Corvids use observational learning to pilfer caches and often re-cache hoards when potential thieves are preoccupied [60, 79, 80, 109]. Re-caching can also serve other functions, such as updating cache-site memory [110-112], redistributing concentrated caches [97, 107, 110, 113], and monitoring the condition of perishable food items [111]. Burns and Van Horik [42] found that food re-caching is a common feature of robin food hoarding behaviour. Both sexes frequently re-cache prey that they had previously hoarded and also re-cache food that was previously hoarded by their mate. Most interestingly, rates of retrieval and pilferage differ between the sexes. Female-made caches are re-cached at the same rate by females as by pilfering males, whereas male-made caches were more frequently re-cached by pilfering females than retrieved by males [42].

Re-caching in robins may serve as a means to reduce cache theft. However, the re-caching of pilfered prey indicates that re-caching is unlikely to be a method of reducing cache theft only. Re-caching in robins may play a role in cache pilfering strategies as a means to effectively exploit the caches of others. Pilfered then re-cached prey will be effectively lost to the individual who originally created the cache. Re-caching allows pilferers to continue to efficiently exploit any of their mate's caches after they become satiated. As a result, robins show a variety of strategies to protect their own caches, as well as steal caches made by their mates, which ultimately distributes food equally between the sexes.

4. WHAT MAKES THE NEW ZEALAND ROBIN UNIQUE? COMPARISON WITH OTHER SYSTEMS

The food hoarding behaviour and natural history of the New Zealand robin is quite different to that of other food hoarding systems. A vast number of experimental and observational studies have been carried out on food-storing mammals, birds and arthropods with the aim of elucidating the conditions that have influenced the evolution of food hoarding behaviour. There is extensive literature available on the food-caching patterns of many bird species. In order to orientate the reader and to distinguish the New Zealand robins' position within the wide variety of food-hoarding systems, this last section will highlight the most important similarities and differences with other, more widely known avian food-hoarding systems.

The diets of nutcrackers and pinyon jays (*Corvidae*), woodpeckers (*Picidae*), nuthatches (*Sittidae*), chickadees, tits and titmice (*Paridae*) all contain seeds and nuts. This is a marked distinction from the robin diet and has important consequences. Seeds are adapted to persist for long periods in a dormant state and are generally resistant to microbial attack [54, 114]. Therefore, seeds take a much longer time to decay and become unpalatable compared to insectivorous prey. Such longevity provides grainivorous food storers with the potential to survive much longer periods of food scarcity by relying on stored food.

Seeds often show a marked seasonal cycle of availability. In higher latitudes there is usually a late summer peak in seed production [38]. Seed hoarding birds often store massive quantities of maturing seeds and nuts in autumn [37, 115-122]. For nutcrackers, pinyon jays, and some chickadees, tits and titmice, seeds stored in autumn form a significant part of their winter diet. It seems likely that the evolution of food hoarding allowed these birds to survive in their habitats during winter. Woodpeckers and nuthatches rely on stored seeds and nuts to provide a consistent supply of resources over shorter periods of food shortage and seed storing enables woodpeckers and nuthatches to maintain their winter territories [90, 117]. In these food hoarding systems the longevity of seeds has enabled the evolution of a long-term food storing behaviour, vastly different to that of the insectivorous robin.

Robin food hoarding behaviour may be more comparable to carrion hoarding by mammalian predators, such as leopards (*Panthera pardus*), than to seed hoarding songbirds in the Northern hemisphere. Leopards drag the carcasses of dead prey into the upper branches of trees for storage. It is widely documented that leopards have a catholic diet, are highly opportunistic [123-125], and Radloff and du Toit [125] found that both males and females can kill prey over three times their bodyweight. Leopards also use caves to store prey. De Ruiter and Berger [124] proposed that caves were the preferred cache site for leopards as the extreme slope of the interiors of caves allowed storage of much larger carcasses than could be transported up into a tree, while also being inaccessible to hyenas. Bailey [126] found that 5-10% of kills are lost to lions and spotted hyenas (*Crocuta crocuta*). Massive prey items, the threat of kleptoparasitism by scavengers, and the inability of lone leopards to defend carcasses may be the driving force behind the evolution of leopard food storage behaviour. New Zealand robins also catch prey that are much larger than themselves. These prey items are dismembered and cached in the forest canopy in sites that would have been highly inaccessible to their many flightless, ground-dwelling animals in prehistoric New Zealand (see section 2). The motivations for caching in robins appear to be quite similar to that of the leopards and they often appear to move prey items up off the forest floor with an air of urgency.

The closest avian behavioural counterpart to New Zealand robins may be the Northwestern crow (*Corvus caurinus*). Both species incorporate food hoarding into their daily cycle of foraging and feeding. Several other avian species display diurnal patterns of food storing intensity, such as some raptors [127-129] and white-breasted nuthatches (*Sitta carolinensis*) [130]., which may help maintain optimal body mass [129, 130]. While energy management may contribute to the pattern of food use in robins [14], a major stimulus for caching in both Northwestern crows and robins is a temporary super-abundance of food. Northwestern crows forage extensively in the intertidal zone [78]. Due to a paucity of food at high tide they are faced with regular periods of food scarcity that are determined by the lunar cycle. The intertidal zone, although exposed only for short periods of time contains very high densities of prey (181clams/m^2 [131]). Northwestern crows commonly cache clams and crabs which the crows capture during low tide[78]. Similarly, many of the invertebrates consumed by New Zealand robins are larger than are found elsewhere in the world and therefore represent an overabundance of food available over a short period of time. New Zealand robins most commonly cache giant earthworms, which are frequently longer than robins [14, 22]. An 200mm long earthworm represents a vast excess of food for the robin and food caching likely helps birds take advantage of this temporary super-abundance of prey. Cache duration in Northwestern crows and robins is similar. The prey of both species rapidly decay and become unpalatable [14, 78]. Thus caches do not persist for long periods and are usually recovered within days [14] (Verbeek 1982 pers. obs. in [78]).

Several aspects of food hoarding in New Zealand robins is similar to that of true shrikes (*Laniidae*), which impale prey on sharp objects within their territories. For shrikes, impaled prey is easier to feed upon, as well as functioning as a cache. Males hold permanent territories and regularly cache small birds, reptiles and mammals. Females are not permanent residents and do not create caches. Cache sizes increase before the breeding season and decline sharply when males feed prey to mate and young (see [132]). Male shrikes often increase the size and visual effect of caches by impaling inedible objects such as rags, snail and eggshells, fæcal sacs or even bread crusts close by, which form as a type of prenuptial display [133-136]. Yosef and Pinshow [132] experimentally manipulated the cache size of male shrikes and found that cache size affects mate selection by females. Males with augmented caches mated earlier and sired more young than controls, and males with caches removed remained unpaired and deserted their territories.

Male New Zealand robins perform a similar prenuptial feeding display, which we have termed the 'food acquisition display'. During the breeding season, males repeatedly pick up as many mealworms as possible in their bill and then drop all the mealworms to the forest floor. This behaviour is often repeated for periods of longer than 5 minutes and has been recorded by previous researchers [16, 17]. It is only ever performed by males and occurs mostly in the breeding season. By continually picking-up and dropping prey, males may be signalling the quality of their territory or their ability to acquire prey [17]. Mating often occurs immediately after males perform the display. After mating, males feed females the mealworms used in the display, which she either consumes or caches. Similarly, the theft of male-made caches by females increases through winter and into spring. In late spring, at the onset of the breeding season, the majority of male-made caches are 'stolen' by females. Although males are very aggressive towards females in winter, they may allow females greater access to their caches in spring as a form of indirect nuptial feeding. As mentioned previously, additional experiments are clearly needed to directly quantify changes in rates of cache theft by females as winter turns to spring.

Finally, New Zealand robins exhibit behaviour that suggests that they are capable of complex cognitive tasks comparable to that documented in corvids, parrots and primates [137]. For example, Bednekoff and Balda [138, 139] documented observational spatial memory in food hoarding corvids, which might also occur in robins. They placed a pair of corvids in an experimental enclosure with many potential cache sites. One bird was allowed to cache while the other was allowed to observe from within a cage. Pinyon jays, Clark's nutcrackers and Mexican jays remembered the locations chosen by conspecifics to cache seeds. The typical social environment of New Zealand robins is quite similar to the design in this experiment. Pairs live together in territories, both sexes keenly observe their partner caching. Both sexes often depart from cache sites to conduct other activities before returning to retrieve caches or pilfer from their mate's caches. When robins retrieve caches they typically fly to the exact location of the cache sites, whether retrieving their own cache or pilfering from cache sites they observed their mates make [94]. Remembering the location of caches made by others is strongly suggestive of observational spatial memory, which to date has only been documented in corvids and primates [138-142]. Additional experiments are now needed to formally test this conjecture.

Robins employ sophisticated strategies of cache protection and pilferage, which may imply that they may have evolved sophisticated cognitive tools to enhance the probability of successful cache creation, protection, retrieval and theft [79, 142-145]. A wide variety of experiments can be conducted to investigate the limits of the cognitive ability of New Zealand robins. Recent phylogenetic work suggests that passerine birds are nested within corvida and that both groups originated in eastern Gondwana [4, 146, 147]. Given that New Zealand was the first to separate from Gondwana, future studies may provide unique insight into the evolution and diversification of avian intelligence in this unusual food hoarding bird.

ACKNOWLEDGEMENTS

The authors would like to thank Dr Anders Brodin and Dr Ralph Powlesland for their comments on this manuscript.

REFERENCES

[1]   Sibley, C.G. and B.L. Monroe, (1990).*Distribution and Taxonomy of the Birds of the World*. New Haven, CT: Yale University Press.

[2]   Sibley, C.G. and J.E. Ahlquist, (1990).*Phylogeny and Classification of Birds: A Study in Molecular Evolution.*, New Haven, CT: Yale University Press.

[3]   Barker, F.K., G.F. Barrowclough, and J.G. Groth, (2002).A phylogenetic hypothesis for passerine birds: taxonomic and biogeographic implications of an analysis of nuclear DNA sequence data. *Proceedings of the Royal Society London B. 269*: p. 295-308.

[4]   Ericson, P.G.P., L. Christidis, and A. Cooper, (2002).A Gondwanan origin of passerine birds supported by DNA sequences of the endemic New Zealand wrens. *Proc. R. Soc. Lond. B. 269*: p. 235-241.

[5]   Fleming, C.A., (1950).New Zealand flycatchers of the genus *Petroica* Swainson (Aves), Parts I and II. *Trans. Roy. Soc. NZ. 78*: p. 14-47, 126-160.

[6]   Miller, H.C. and D.M. Lambert, (2006).A molecular phylogeny of New Zealand's *Petroica* (Aves: Petroicidae) species based on mitochondrial DNA sequences. *Molecular Phylogenetics and Evolution. 40*: p. 844-855.

[7] Heather, B.D. and H.A. Robertson, (2005).*The field guide to the birds of New Zealand.* Auckland: Penguin Books Ltd.

[8] Higgins, P.J. and J.M. Peter, (2002).*Handbook of Australian, New Zealand and Antarctic birds. Volume 6: pardalotes to shrike-thrushes.*, Victoria: Oxford University Press.

[9] Armstrong, D.P., (2001).Sexing North Island robins (*Petroica australis longipes*) from morphometrics and plumage. *Notornis. 48*: p. 76-80.

[10] Berggren, A., D.P. Armstrong, and R.M. Lewis, (2004).Delayed plumage maturation increases overwinter survival in North Island robins. *Proc. R. Soc. Lond. B. 271*: p. 2123-2130.

[11] Armstrong, D.P., J.G. Ewen, and W.J. Dimond, (2000).Breeding biology of North Island robins (*Petroica australis longipes*) on Tiritiri Matangi Island, Hauraki Gulf, New Zealand. *Notornis. 47*: p. 106-118.

[12] Powlesland, R.G., (1983).Breeding and mortality of the South Island robin in Kowhai Bush, Kaikoura. *Notornis. 30*: p. 265-282.

[13] Alexander, L., et al., (2005).An experimental evaluation of food hoarding by North Island robins (*Petroica australis longipes*). *Notornis. 52*: p. 138-142.

[14] Powlesland, R.G., (1980).Food storing behaviour of the South Island robin. *Mauri Ora. 8*: p. 11-20.

[15] Flack, J.A.D., (1976).The use of frontal spot and crown feathers in inter- and intraspecific display by the South Island robin, *Petroica australis australis. Notornis. 23*: p. 90-105.

[16] Moncrieff, P., (1931).Notes on the South Island robin. *Emu. 332*: p. 52-58.

[17] Soper, M.F., (1976).*New Zealand Birds.* Christchurch: Whitchoulls Publishers.

[18] Powlesland, R.G., (Personal communication).

[19] Burns, K.C. and J. Steer, (2006).Dominance rank influences food hoarding in New Zealand robins *Petroica australis. Ibis. 148*: p. 266-272.

[20] Powlesland, R.G., (1981).Comparison of time-budgets for mainland and Outer Chetwode Island populations of adult male South Island robins. *New Zealand Journal of Ecology. 4*: p. 98-105.

[21] Daugherty, C.H., G.W. Gibbs, and R.A. Hitchmough, (1993).Mega-island or micro-continent? New Zealand and its fauna. *Trends in Ecology and Evolution. 8*: p. 437-442.

[22] Powlesland, R.G., (1981).The foraging behaviour of the South Island robin. *Notornis. 28*: p. 89-102.

[23] Lee, K.E., (1959).*The earthworm fauna of New Zealand.* Wellington: Government Printer.

[24] Jewell, T. and P.D. Brock, (2002).A review of the New Zealand stick insects: New genera and synonymy, keys and a catalogue. *Journal of Orthoptera Research. 11*: p. 189-197.

[25] Gibbs, G.W., (1998).*New Zealand weta.*, Auckland: Reed Publishing.

[26] Flack, J.A.D., (1973).Robin research - a progress report. *Wildelife - A Review. 4*: p. 28-36.

[27] Cooper, R.A. and P.R. Millener, (1993).The New Zealand biota: historical background and new research. *Trends in Ecology and Evolution. 8*: p. 429-433.

[28] Blackwell, G.L., (2005).Another world: the composition and consequences of the introduced mammal fauna of New Zealand. *Australian Zoologist. 33*: p. 108-118.

[29] Gibbs, G.W., (2006).*Ghosts of Gondwana: The history of life in New Zealand*. Nelson, NZ: Craig Potton Publishing.

[30] Duthie, C., G.W. Gibbs, and K.C. Burns, (2006).Seed dispersal by weta. *Science. 311*: p. 1575.

[31] McNabb, B.K., (1994).Energy conservation and the evolution of flightlessness in birds. . *American Naturalist. 144*: p. 628-642.

[32] Carlquist, S., (1965).*Island life*. New York: Natural History Press.

[33] Bell, B.D., (1990).Recent avifaunal changes and the history of ornithology in New Zealand. *Acta XX Congressus Internationalis Orthithologici*: p. 193-230.

[34] Diamond, J.M. and C.R. Veitch, (1981).Extinctions and introductions in the New Zealand avifauna: Cause and effect? *Science. 211*: p. 499-501.

[35] Robertson, C.J.R., et al., (2007).*Atlas of bird distribution in New Zealand 1999-2004*. Wellington, N.Z.: Ornithological Society of New Zealand.

[36] BirdLife International, *Petroica australis*, in *2006 IUCN Red List of Threatened Species*, IUCN 2006, Editor. 2004: from http://www.iucnredlist.org.

[37] Haftorn, S., (1953).Contribution to the food biology of tits especially about storing of surplus food. Part 1. The crested tit (*Parus cristatus* L.). *Det Kgl Norske Videnskabers Selskabs Skrifter. 1953*(4): p. 1-122.

[38] Vander Wall, S.B., (1990).*Food Hoarding in Animals*. Chicago: University of Chicago Press.

[39] Steer, J. and J. Van Horik, (2006).North Island robin (*Petroica australis longipes*) food caches are stolen by stichbirds (*Notiomystis cincta*) and bellbirds (*Anthornis melanura*). *Notornis. 53*: p. 315-316.

[40] Powlesland, R.G., *A time-budget study of the South Island robin (Petroica a. australis) at Kowhai Bush, Kaikoura*. 1980, University of Canterbury: Christchurch New Zealand.

[41] Barr, K., et al., (1996).Impacts of introduced common wasps (*Vespula vulgaris*) on experimentally placed mealworms in a New Zealand beech forest. *Oecologia. 105*: p. 266-270.

[42] Burns, K.C. and J. Van Horik, (2007).Sexual differences in food re-caching by New Zealand robins *Petroica australis*. *Journal of Avian Biology. 38*: p. 394-398.

[43] Atkinson, I.A.E. and P.R. Millener, (1990).An ornithological glimps into New Zealand's pre-human past. *Acta XX Congressus Internationalis Orthithologici*: p. 129-192.

[44] Miskelly, C., R. Empson, and K. Wright, (2005).Forest bird recolonising Wellington. *Notornis. 52*: p. 21-26.

[45] Steer, J., *Seasonality of food use and caching in New Zealand robins (Petroica australis)*, in *Biological Sciences*. 2006, Victoria University of Wellington: Wellington, New Zealand.

[46] MacDougall-Shackleton, S.A., et al., (2003).Photoperiodic regulation of food storing and hippocampus volume in black-capped chickadees, *Poecile atricapillus*. *Animal Behaviour. 65*: p. 805-812.

[47] Lack, D., (1968).*Ecological adaptations for breeding in birds*. London, UK: Methuen.

[48] Sherry, D.F., (1989).Food storing in the paridae *Wilson Bulletin. 101*: p. 289-304.

[49] Pravosudov, V.V., (2006).On seasonality in food-storing behaviour in parids: do we know the whole story? *Animal Behaviour. 71*: p. 1455-1460.

[50]   Steer, J. and K.C. Burns, (2008).Seasonal variation in sexual conflict, cooperation and selfish hoarding in a monogamous songbird. *Behavioral Ecology and Sociobiology. 62*: p. 1175-1183.

[51]   McNamara, J.M., A.I. Houston, and J.R. Krebs, (1990).Why hoard? The economics of food hoarding in tit, *Parus spp. Behavioral Ecology. 1*: p. 12-23.

[52]   Lucas, J.R. and D.L. Zielinski, (1998).Seasonal variation in the effect of cache pilferage on cache and body mass regulation in Carolina Chickadees: what are the trade-offs? *Behavioral Ecology. 9*: p. 193-200.

[53]   Pravosudov, V.V. and T.C. Grubb, (1998).Management of fat reserves in Tufted Titmice (*Parus bicolor*): evidence against a trade-off with food hoards. *Behavioral Ecology and Sociobiology. 42*: p. 57-62.

[54]   Gendron, R.P. and O.J. Reichman, (1995).Food perishability and inventory management: a comparison of three caching strategies. *American Naturalist. 145*: p. 948-968.

[55]   Witter, M.S., I.C. Cuthill, and R.H.C. Bonser, (1994).Experimental investigations on mass-dependant predation risk in the European Starling, *Sturnus vulgaris. Animal Behaviour. 48*: p. 201-222.

[56]   Brodin, A., (2001).Mass-dependant predation and metabolic expenditure in wintering birds; is there a trade-off between different forms of predation? . *Animal Behaviour. 62*: p. 993-999.

[57]   Gosler, A.G., (2002).Strategy and constraint in the winter fattening response to temperature in the great tit *Parus major. Journal of Animal Ecology. 71*: p. 771-779.

[58]   Pravosudov, V.V., et al., (1999).Social dominance and energy reserves in wintering woodland birds. *The Condor. 101*: p. 880-884.

[59]   Brodin, A., (2007).Theoretical models of adaptive energy management in small wintering birds. *Philos Trans R Soc Lond B. 362*: p. 1857-1871.

[60]   Emery, N.J. and N.S. Clayton, (2001).Effects of experience and social context on prospective caching strategies by Scrub Jays. *Nature. 414*: p. 443-446.

[61]   Vander Wall, S.B. and S.H. Jenkins, (2003).Reciprocal pilferage and the evolution of food-hoarding behaviour. *Behavioral Ecology. 14*: p. 656-667.

[62]   Lahti, K. and S. Rytkönen, (1996).Presence of conspecifics, time of day and age affect Willow Tit food hoaring. *Animal Behaviour. 52*: p. 631-636.

[63]   Brodin, A., K. Lundberg, and C.C. W., (2001).The effect of dominance on food hoarding: a game theoretical model. *The American Naturalist. 157*(66-75).

[64]   Lahti, K., (1998).Social dominance and survival in flocking passerine birds: a review with an emphasis on the Willow Tit, *Parus montanis. Ornis Fennica. 75*: p. 1-17.

[65]   Smith, R.D., G.D. Ruxton, and W. Cresswell, (2002).Do kleptoparasites reduce their own foragin effort in order to detect kleptoparasite opportunities? An empirical test of a key assumption of kleptoparasite models. *Oikos. 97*: p. 205-212.

[66]   Lundberg, K. and A. Brodin, (2003).The effect of dominance rank on fat deposition and food hoarding in the Willow Tit *Parus montanus* - an experimental test. *Ibis. 145*: p. 78-82.

[67]   Andersson, M. and J.R. Krebs, (1978).On the evolution of hoarding behaviour. *Animal Behaviour. 26*: p. 707-711.

[68]   Stapanian, M.A. and C.C. Smith, (1978).A model for seed scatter-hoarding: coevolution of fox squirrels and black walnuts. *Ecology. 59*: p. 884-896.

[69] Smulders, T.V., (1998).A game theoretical model of the evolution of food hoarding:applications to the *Paridae. American Naturalist. 151*: p. 356-366.

[70] Brodin, A. and J.B. Ekman, (1994).Benefits of food hoarding. *Nature. 372*: p. 510.

[71] Ekman, J.B., A. Brodin, and B. Sklepkovych, (1995).Selfish long-term benefits of hoarding in the Siberian jay. *Behavioral Ecology. 7*: p. 37-43.

[72] Vander Wall, S.B., et al., (2006).Joshua tree (Yucca brevifolia) seeds are dispersed by seed-caching rodents. *Ecoscience. 13*(4): p. 539-543.

[73] Dally, J.M., N.J. Emery, and N.S. Clayton, (2006).Food-caching western scrub-jays keep track of who was watching when. *Science. 312*(5780): p. 1662-1665.

[74] Briggs, J.S. and S.B. Vander Wall, (2004).Substrate type affects caching and pilferage of pine seeds by chipmunks. *Behavioral Ecology. 15*: p. 666-672.

[75] Dally, J.M., N.J. Emery, and N.S. Clayton, (2004).Cache protection strategies in western scrub-jays (*Aphelocoma californica*): hiding food in the shade. *Proc. R. Soc. Lond. B. 271*: p. S387-S390.

[76] Burnell, K.L. and D.F. Tomback, (1985).Stellar's jays steal grey jay caches: field and laboratory observations. *Auk. 102*: p. 417-419.

[77] Carrascal, L.M. and E. Moreno, (1993).Food caching versus immediate consumption in the nuthatch: ehe effect of social context. *Ardea. 81*: p. 135-141.

[78] James, P.C. and N.A.M. Verbeek, (1983).The food storage behaviour of the northwestern crow. *Behaviour. 85*: p. 276-291.

[79] Dally, J.M., N.J. Emery, and N.S. Clayton, (2005).Cache protection strategies by western scrub-jays, Aphelocoma californica: implications for social cognition. *Animal Behaviour. 70*: p. 1251-1263.

[80] Emery, N.J., J.M. Dally, and N.S. Clayton, (2004).Western scrub-jays (*Aphelocoma californica*) use cognitive strategies to protect their caches from theiving conspecifics. *Animal Cognition. 7*: p. 37-43.

[81] Jokinen, S. and J. Suhonen, (1995).Food caching by willow and crested tits: a test of scatter hoarding models. *Ecology. 76*: p. 892-898.

[82] Lahti, K., et al., (1998).Social influences on food caching in willow tits: a field experiment. *Behavioral Ecology. 9*: p. 122-129.

[83] Alatalo, R.V. and A. Carlson, (1987).Hoarding-site selection of the willow tit *Parus montanus* in the presence of the Siberian tit *Parus cinclus. Ornis Fennica. 64*: p. 1-9.

[84] Moreno, J., A. Lundberg, and A. Carlson, (1981).Hoarding of individual nuthatches *Sitta europea* and march tits *Parus palustris. Holarctic Ecology. 4*: p. 263-269.

[85] Woodrey, M.S., (1990).Economics of caching versus immediate consumption by white-breasted nuthatches: the effect of handling time. *Condor. 92*: p. 621-624.

[86] Brotons, L., (2000).Individual food hoarding decisions in a non-territorial coal tit population: the role of social context. *Animal Behaviour. 60*: p. 395-402.

[87] Clarke, M.F. and D.L. Kramer, (1994).Scatter-hoarding by a larder-hoarding rodentl intraspecific variation in the hoarding behaviour of the eastern chipmunk (*Tamias striatus*). *Animal Behaviour. 48*: p. 299-308.

[88] Clarkson, K., et al., (1986).Density dependence and magpie food hoarding. *Journal of Animal Ecology. 55*: p. 111-121.

[89] Spritzer, M.D. and D. Brazeau, (2003).Direct vs. indirect benefits of caching by grey squirrels (*Sciurus carolinensis*). *Ethology. 109*: p. 559-575.

[90] Koenig, W.D. and R.L. Muume, (1987).*Population Ecology of the Cooperatively Breeding Acorn Woodpecker.*, Princeton, New Jersey: Princeton University Press.

[91] Novakowski, N.S., (1967).The winter bioenergenics of a beaver population in northern latitudes. *Canadian Journal of Zoology. 45*: p. 1107-1118.

[92] Boutin, S., K.W. Larsen, and D. Berteaux, (2000).Anticipatory parental care: acquiring resources for offspring prior to conception. *Proc. R. Soc. Lond. B. 267*: p. 2081-2085.

[93] Dally, J.M., N.J. Emery, and N.S. Clayton, (2005).The social suppression of caching in western scrub-jays (*Aphelocoma californica*). *Behaviour. 142*: p. 961-977.

[94] Van Horik, J. and K.C. Burns, (2007).Cache spacing patterns and reciprocal cache theft in New Zealand robins. *Animal Behaviour. 73*: p. 1043-1049.

[95] Barnea, A. and F. Nottebohn, (1995).Patterns of food storing by black-capped chickadees suggest a mnemonic hypothesis. *Animal Behaviour. 49*: p. 1161-1176.

[96] Petit, D.R., L.J. Petit, and K.E. Petit, (1989).Winter caching ecology of deciduous woodland birds and adaptations for protection of stored food. *Condor. 91*: p. 766-776.

[97] Waite, T.A. and J.D. Reeve, (1992).Grey jay scatter hoarding behaviour, rate maximisation, and the effect of local cache density. *Ornis Scandinavica. 23*: p. 175-182.

[98] Daly, M., et al., (1992).Scattern hoarding by kangaroo rats (*Dipodomys merriami*) and pilferage from their caches. *Behavioral Ecology. 3*: p. 102-111.

[99] Sherry, D.F., M. Avery, and S. A., (1982).The spacing of stored food by marsh tits. *Zeitschrift für Tierpsychologie. 58*: p. 153-162.

[100] Tamura, N., Y. Hashimoto, and F. Hayashi, (1999).Optimal distances for squirrels to transport and hoard walnuts. *Animal Behaviour. 58*: p. 635-642.

[101] Waite, T.A., (1988).A field test of density dependent survival of simulated grey jay caches. *Condor. 90*: p. 247-249.

[102] Baker, M.C., M.D. Mantych, and R.J. Shelden, (1990).Social dominance, food caching and recovery by black-capped chickadees (*Parus atricapillus*): is there a cheater strategy? *Ornis Scandinavica. 21*: p. 293-295.

[103] Brodin, A., (1994).Separation of caches between individual willow tits hoarding under natural conditions. *Animal Behaviour. 47*: p. 1031-1035.

[104] Ekman, J.B., (1979).Coherence, comosition and territories of winter social groups of the willow tit (*Parus montanus*) and thr crested tit (*P. cristatis*). *Ornis Scandinavica. 10*: p. 56-68.

[105] Lens, L., F. Adriaensen, and A.A. Dhont, (1994).Age-related hoarding strategies in the crested tit *Parus cristatus*: should the cost of subordination be re-assessed. *Journal of Animal Ecology. 63*: p. 749-755.

[106] Smith, S.M., (1991).*The Black-Capped Chickadee: Behavioural ecology and Natural History*. New York: Cornell University Press.

[107] Brodin, A., (1993).Low rates of loss of willow tit caches may increase adaptiveness of long-term hoarding. *Auk. 110*: p. 642-645.

[108] Ekman, J.B., (1990).Alliances in winter flocks of willow tits; effects of rank on survival and reproductive success in male-female associations. *Behavioral Ecology and Sociobiology. 26*: p. 239-245.

[109] Preston, S.E. and L.F. Jacobs, (2001).Conspecific pilferage but not presence affects merriam's kangaroo rat cache strategy. *Behavioral Ecology. 12*: p. 517-523.

[110] Brodin, A., (1992).Cache dispersion affects retrieval time in hoarding willow tits. *Ornis Scandinavica. 23*: p. 7-12.

[111] DeGrange, A.R., et al., (1989).Acorn harvesting by Florida scrub jays. *Ecology. 70*: p. 348-356.

[112] Grubb, T.C. and V.V. Pravosudov, (1994).Toward a general theory of energy management in wintering birds. *Journal of Avian Biology. 25*: p. 255-260.

[113] Waite, T.A. and J.D. Reeve, (1995).Source use decisions by hoarding grey jays: effects of local cache density and food value. *Journal of Avian Biology. 26*: p. 59-66.

[114] Smith, C.C. and O.J. Reichman, (1984).The evolution of food caching by birds and mammals. *Annual Review of Ecology and Systematics. 15*: p. 329-51.

[115] Balda, R.P. and G.C. Bateman, (1971).Flocking and annual cycle of the piñon jay, *Gymnorhinus cyanocephalus. Condor. 73*: p. 287-302.

[116] Brodin, A., (1994).The role of naturally stored food supplies in the winter diet of the boreal willow tit *Parus montanus. Ornis Svecica. 4*: p. 31-40.

[117] Enoksson, B., (1987).Local movements in the nuthatch *Sitta europeae Acta Reg. Soc. Sci. Litt. Gothoburgensis Zool. 14*: p. 36-47.

[118] Haftorn, S., (1956).Contribution to the food biology of tits especially about storing of surplus food. Part 3. The willow-tit (*Parus atricapillus* L.). *Det Kgl Norske Videnskabers Selskabs Skrifter. 1956*(3): p. 1-78.

[119] Haftorn, S., (1956).Contribution to the food biology of tits especially about storing of surplus food. Part 2. The coal-tit (*Parus alter* L.). *Det Kgl Norske Videnskabers Selskabs Skrifter. 1956*(2): p. 1-52.

[120] Haftorn, S., (1956).Contributions to the food biology of tits especially about storing of surplus food. Part 4. A comparitive analysis of *Parus atricapillis* L., *P. cristatus* . and *P. ater* L. *Det Kgl Norske Videnskabers Selskabs Skrifter. 1956*(4): p. 1-54.

[121] Moskovits, D., (1978).Winter territorial and foraging behaviour of red-headed woodpeckers in Florida. *Wilson Bulletin. 90*: p. 512-535.

[122] Tomback, D.F., (1982).Dispersal of whitebark pine seeds by Clark's nutcracker: A mutualism hypothesis. *Journal of Animal Ecology. 51*: p. 451-467.

[123] Balme, G., L. Hunter, and R. Slotow, (2006).Feeding habitat selection by hunting leopards *Panthera pardus* in a woodland savanna: prey catchability versus abundance. *Animal Behaviour. 74*: p. 589-598.

[124] de Ruiter, D.J. and L.R. Berger, (2001).Leopard (*Panthera pardus* Linneaus) cave caching related to anti-theft behaviour in the John Nash Nature Reserve, South Africa. *African Journal of Ecology. 39*: p. 396-398.

[125] Radloff, F.G. and J.T. du Toit, (2004).Large predators and their prey in a southern African savanna: a predator's size determines its prey size range. *Journal of Animal Ecology. 73*: p. 410-423.

[126] Bailey, T.N., (2005).*The African Leopard: Ecology and Behaviour of a Solitary Felid.*, 2nd edn. Caldwell, New Jersey: Blackburn Press.

[127] Collopy, M.W., (1977).Food caching behaviour by female American kestrels in winter. *Condor. 79*: p. 63-68.

[128] Oliphant, L.W. and W.J.P. Thompson, (1976).Food caching behaviour in Richardson's merlin. *Canadian Field-Naturalist. 90*: p. 364-365.

[129] Rijnsdorp, A., S. Daan, and C. Dijkstra, (1981).Hunting in the kestral, *Falco tinnunculus*, and the adaptive significance of daily habits. *Oecologia. 50*: p. 391-406.

[130] Waite, T.A. and T.C. Grubb, (1988).Diurnal caching rhythm in captive white-brested nuthatches (*Sitta carolinensis*). *Ornis Scandinavica. 19*: p. 68-70.

[131] Butler, R.W., *The breeding ecology and social organisation of the Northwestern crow on Mitlenatch Island B.C.* 1979, Simon Fraser University.

[132] Yosef, R. and B. Pinshow, (1989).Cache size in shrikes influences female mate choice and reproductive success. *Auk. 106*: p. 418-421.

[133] Durango, S., (1956).Territory in the Red-backed Shrike *Lanius collurio. Ibis. 98*: p. 476-484.

[134] Leonard, D.L., (1992).Monthly variation in a Loggerhead Shrike cache in central Florida. *Florida Field Naturalist. 20*: p. 104-107.

[135] Yosef, R. and B. Pinshow, (1988).Polygyny in the Northern Shrike, *Lanius excubitor*, in Isreal. *Auk. 105*: p. 581-582.

[136] Yosef, R. and B. Pinshow, (2005).Impaling in true shrikes (*Laniidae*): A behavioural and ontogentic perspective. *Behavioural Processes. 69*: p. 363-367.

[137] Baker, M.C., et al., (1988).Evidence against observational learning in storage and recovery of seeds by black-capped chickadees. *Auk. 105*: p. 492-497.

[138] Bednekoff, P.A. and R.P. Balda, (1996).Social caching and observatoinal spatial memory in pinyon jays. *Behaviour. 133*: p. 807-826.

[139] Bednekoff, P.A. and R.P. Balda, (1996).Observational spatial memory in Clark's nutcrackers and Mexican jays. *Animal Behaviour. 52*: p. 833-839.

[140] Clayton, N.S. and A. Dickinson, (1998).Episodic-like memory during cache recovery by western scrub jays. *Nature. 395*: p. 272-278.

[141] Clayton, N.S., et al., (2001).Elements of episodic-like memory in animals. *Philosophical Transactions: Biological Sciences. 356*: p. 1483-1491.

[142] Emery, N.J. and N.S. Clayton, (2004).The mentality of crows: Convergent evolution of intelligence in corvids and apes. *Science. 306*: p. 1903-1907.

[143] Correia, S.P.C., A. Dickinson, and N.S. Clayton, (2007).Western Scrub-jays anticipate future needs independantly of their current motivational state. *Current Biology. 17*: p. 856-861.

[144] Raby, C.R., et al., (2007).Planning for the future by western scrub-jays. *Nature. 445*(7130): p. 919-921.

[145] Roberts, W.A., (2007).Mental time travel: animals anticipate the future. *Current Biology. 17*: p. R418-R420.

[146] Edwards, S.V. and W.E. Boles, (2002).Out of gondwana: the origin of passerine birds. *TRENDS in Ecology and Evolution. 17*: p. 347-349.

[147] Ericson, P.G.P., M. Irestedt, and U.S. Johansson, (2003).Evolution, biogeography and patterns of diversification in passerine birds. *Journal of Avian Biology. 34*: p. 3-15.

Reviewed by: Dr. Anders Brodin and Dr. Ralph Powlesland

In: Animal Behavior: New Research
Editors: E. A. Weber, L. H. Krause

ISBN: 978-1-60456-782-3
© 2008 Nova Science Publishers, Inc.

Chapter 8

DISCRIMINATIVE LEARNING, LEARNING GENERALIZATION AND MASKING TESTS AS THREE STRATEGIES TO ASSESS OLFACTORY DISCRIMINATION

Julien Colomb

Gènes et Dynamique des Systèmes de Mémoire, CNRS UMR 7637, ESPCI,
10 rue Vauquelin, 75005 Paris, France

ABSTRACT

While our understanding of olfactory information coding in the brain is rapidly rising, knowledge of how different odors are perceived and discriminated in the brain remains sparse and speculative. One of the primary reasons for this is the difficulty in assessing if an animal can discriminate odor A from odor B. I review here three different strategies envisaged so far and their caveats: discriminative learning, learning generalization and masking tests.

Discriminative learning involves presenting an odor A with a reinforcer and B alone. The animal is then tested for its preference for A over B. The result depends both on the learning and memory abilities of the animal and on the similarity of the odor used (animals cannot perform well if the A and B are too similar). Generalization learning involves presenting only A either with the reinforcer (associative learning generalization) or without (sensitivity to A will decrease, and cross-adaptation can be tested). By testing the behavioral response toward B, the generalization from A to B is measured. The magnitude of this generalization was shown to be dependent on the similarity of A and B. These paradigms have been used with some success, although their dependence on learning tasks complicates interpretations. A third strategy, called a masking test, has not been extensively used so far, despite its simplicity and efficacy. Here, the chemotaxis of animals toward A is measured in the background of a saturated concentration of B. Although the biological mechanisms involved in this test are debated (olfactory adaptation to B was postulated to be crucial), it seems independent of learning mechanisms and may thus help to dissect the molecular and cellular bases of olfactory discrimination in the future.

Detailed information about neural representations of olfactory signals in the brain of vertebrates and invertebrates has been gathered over recent decades. It has become clear that those two evolutionarily different systems share the same functional architecture (for a review see Korsching, 2002; Lin et al, 2007; Vosshall & Stocker, 2007): each receptor neuron expresses one functional type of odorant receptor protein, and all receptors of the same type target the same glomerulus in the primary olfactory center. Olfactory representation is transformed at this layer thanks to inhibitory and exitatory interneurons (Wilson et al, 2004). The information is then transmitted to higher brain centers via neurons receiving monosynaptic input of receptor neurons from a unique glomerulus, where odor recognition may occur. It has become clear that the different properties of the sensed chemicals (functional groups, carbon chain length, ...) are encoded by the olfactory system, but we still understand little of the way in which odors are actually perceived and discriminated in the brain.

This is an important neurobiological question, since the olfactory system has both to recognize differences and similarities in odor blends. For example, flies need to detect fermented banana irrespective of the exact fermentation state, and they have to recognize it as the same despite differences in odor concentration. In order to understand how the olfactory system deals with this question, behavioural measures of olfactory discrimination are necessary. Such assays would allow us to link neural representation to odor perception. A simple test of the preference of animals for two odors cannot be used as an estimation, since a positive chemotaxis does not indicate that odors are perceived as different (they can be perceived as the same odor at different concentrations) and an absence of choice does not imply that animals cannot discriminate the odors (for instance, in olfactory learning studies, the concentration of two discriminated odors are manipulated to avoid any naive preference for one or the other odor (Scherer et al, 2003)). Currently, three strategies have been developed to test olfactory discrimination in animal models: discriminative learning, learning generalisation (and its variant cross-adaptation also called cross-habituation), and olfactory masking tests (Table 1). Strategies in psychophysics (human studies, for a review, see Wise et al, 2000) and a fourth strategy involving a matching-to-sample paradigm developed recently in rats (Pena et al, 2006) are not included in this review. Here, I will present these three different strategies, their advantages and caveats, and illustrate them with some interesting results.

Table 1. The three different stategies to test olfactory discrimination in animals

	Conditioning	Test	Control
Discriminative learning	A+ B/ A B+	A versus B	none
Learning generalization Associative learning	A+	B, C, D, …	naïve B/none
Cross-adaptation	A	B	naïve B
Masking test	none	BA versus A	naïve B

DISCRIMINATIVE LEARNING

In discriminative learning, animals learn to behave differently in response to two tested stimuli; the accuracy or the speed of their response can then be used as a measure of the perceived similarity of the odorants. For instance, one can present an odor A with a reinforcer (for example sugar) and the second odor B alone. By testing the preference of the animal for A toward B, the perceived similarity of A and B can be estimated (the preference of animals for A being maximal for completely discriminated odorants and null for undiscriminated ones). The reciprocal experiment where B is reinforced is usally performed, and a discriminative learning score is calculated from the two experiments. This technique was extensively used in invertebrate studies (Abarca et al, 2002; Borst, 1983; Daly et al, 2001; Sakura et al, 2004; Xia & Tully, 2007) and primate studies (for example see Laska et al, 1999b), but marginaly in rodents (Cleland et al, 2002; Linster et al, 2002; Pavlis et al, 2006), where authors generally preferred to use a go-nogo task: a first odor signals the animal to go forward and be rewarded, a second one signals retraction of its head (for a review see Friedrich et al, 2004). The latter experiment can give information about the time needed by the animal to discriminate odors, which appears to be correlated with the similarity of the odors (as expected from their chemical features, Friedrich et al, 2004).

The major caveat of these studies is that the final measure depends both on discrimination and learning abilities. The salience of odors determines their facility to be learned and varies with their nature and concentration (Pelz et al, 1997). Therefore, the discriminative learning score is dependent upon these variables, which are independent of the similarity of the odors. In addition, when a treatment affects the score, it is difficult to dissociate effects on learning from effects on discrimination. However, by showing that learning is normal with certain (easily discriminated) odor pairs, but not with others (which are perceived as more similar), it is possible to dissociate both effects. As another caveat, animals learn to differentiate the odorants in this paradigm. Indeed, odors that are naively not discriminated (assessed with one of the techniques described below), can be discriminated after such a procedure (Fletcher & Wilson, 2002; Linster et al, 2002; Linster et al, 2001). Interestingly, this is correlated with changes in the neural representation of the odors (Faber et al, 1999). Therefore, one should be aware that an effect on discrimination as tested with this procedure may be linked to an effect on this specific learning.

LEARNING GENERALISATION

Another disadvantage of the discriminative learning paradigm is that each odor pair has to be assessed in a separate experiment. In order to test a multitude of odours simultaneously, one can use a different learning task. Here, only odor A is presented during the learning phase. It is either associated with another stimulus (rewarding or punishing) or presented alone (leading to a non-associative learning called olfactory adaptation: a drop in the sensitivity toward that odorant). When a second odor B is presented during the test phase, the animal may behave according to a generalization mechanism. Generalization allows animals to treat different but similar stimuli as equivalent. Similarity along one or several perceptual dimensions determines the degree of generalization between stimuli (Shepard, 1987). In other words, the extent to which the animal will respond to B depends on the perceived similarity of A and B. Such approaches have been used with different model system such as the bee (Guerrieri et al, 2005; Laska et al, 1999a), the fly (Boyle & Cobb, 2005; Chandra & Singh, 2005; Cobb & Domain, 2000), and the moth (Daly et al, 2001), as well as mammals (Cleland et al, 2002; Fletcher & Wilson, 2002; Linster et al, 2001).

I gathered associative learning generalization and cross adaptation in the same section because they share many properties, although the nature of their effects may be different (Cleland et al, 2002). Though, the two phenomenons may be more similar than expected. Learning is a central mechanism, whereas adaptation was thought to be peripheral (Zufall & Leinders-Zufall, 2000). However, the peripheral nature of olfactory adaptation is currently in debate (Kadohisa & Wilson, 2006; Wilson, 2000). Moreover, the behavior paradigms used to produce olfactory adaptation were shown to lead to associative learning in *Drosophila* (Colomb et al, 2007) and *C.elegans* (Nuttley et al, 2002), meaning that results classed in cross-adaptation studies were actually based on associative learning generalisation (see for example Boyle & Cobb, 2005; Colbert & Bargmann, 1995; Keller & Vosshall, 2007; L'Etoile & Bargmann, 2000).

Using learning generalization paradigms, the discrimination of multiple odor pairs can be tested in one experiment. Once odor A has been learned, one can assess the response toward a series of different odors (four in Guerrieri et al, 2005), or toward different odors in parallel (48 in Laska et al, 1999a). In addition, animals are naive to the second odor in this paradigm - the impact of learning on discriminative ability is thus lower than with the discriminative learning task (Linster et al, 2002; Linster et al, 2001), although it is not null (for a review, see Wilson & Stevenson, 2003). Other caveats exist. First, this paradigm involves a learning procedure, so there are also difficulties to differentiate treatment effects on learning and on discrimination. Moreover, this test has two phases (learning and test), during which discriminative abilities may be differently involved. Indeed, it appeared that different cellular elements are required during the learning and the test phases in order to get normal discrimination (Guerrieri et al, 2005). Finally, a major caveat of this paradigm is that this test measures generalization of a learning and not discrimination per se: even though the two processes measure the relative similarity between odors, the neural and cellular processes involved may be different. In other words, the fact that an animal is generalizing its learning to a dissimilar odor does not mean that he cannot recognize the two odors as different, and vice versa.

MASKING TEST

The third strategy used to test olfactory discrimination does not involve learning. In the masking test, we measure the response of the animal toward an odor, while another saturated odor is present in the background. A decrease in the response in the presence of the background odor is interpreted as a partial discrimination. Proved to be effective in tests of gustatory discriminative abililty (Miyakawa, 1982), the masking test has only marginally been used in research on olfaction (Colomb, 2006; Kelliher et al, 2003; Rodrigues, 1980; Xia & Tully, 2007).

A reason for this dearth of studies is that the mechanism that allows the animal to perform this task is unknown. A simple theory states that different receptors have to be activated in the periphery, but another postulates that olfactory adaptation to the background odor is crucial (Kelliher et al, 2003). Another difficulty comes from the fact that the masking effect appeared to depend on the concentrations of both the mask and the test odors (Rodrigues, 1980). One way to overcome this problem is to choose the concentration of the odors accordingly to the behavioural response toward that odor. For instance, the *Drosophila* larvae response to increasing concentrations of odors follows a sigmoid curve (Cobb, 1999; Colomb et al, 2007). In unpublished experiments, we decided to use the first concentration giving a maximal response as a test odor, and 100 times more concentrated odorant as mask test (Colomb, 2006), which allowed us to compare the effects of different masks on the response to one odorant. Despite these caveats, this technique has the great advantage of being independent of learning mechanisms, and is therefore likely to be used more regularly in the future.

CONCLUSION

Animals can recognize different odor blends as similar or dissimilar, depending on the nature of these blends. The three paradigms presented herein can effectively estimate these perceived similarities. Interestingly, the three methods lead to similar conclusions about the chemical features important for odorant discrimination, independently of the animal model tested. Research on learning generalization in bees (Guerrieri et al, 2005; Laska et al, 1999a), research on Drosophila using cross-adaptation (Cobb & Domain, 2000; Keller & Vosshall, 2007) or masking tests (Colomb, 2006), and research in mammals using discriminative learning and both variants of learning generalisation (Cleland et al, 2002) correlate well with studies in humans (Keller & Vosshall, 2007; Wise et al, 2000). For example, the similarity of odors sharing the same functional group appears linearly correlated to the carbon chain length of the chemicals (Cleland et al, 2002; Cobb & Domain, 2000; Guerrieri et al, 2005; Keller & Vosshall, 2007; Linster et al, 2002; Linster et al, 2001). Moreover, this correlates well with the similarity of the neural representation of these odors in the primary olfactory center (Guerrieri et al, 2005; Keller & Vosshall, 2007; Linster et al, 2002; Linster et al, 2001).

In addition, odor discrimination appears to be asymmetric when measured with learning generalization (Guerrieri et al, 2005) and masking tests (Cobb & Domain, 2000), while such an asymmetry cannot be tested in the discriminative learning paradigm. However, this striking similarity in the two assays might be a coincidence of two different artefacts, and is not linked to odor discrimination per se. While the degree of asymmetry in generalization learning depends on the associated odor and certainly its salience (certain odors generalise more easily than others, Guerrieri et al, 2005), the asymmetry in masking tests was expected to be due to their dependence on receptors activated in the periphery: the more they are activated by the odor, the greater the chance of masking another odor.

Searching for the cellular or molecular correlates of olfactory discrimination appears to be difficult. First, the available tests do not directly quantifiy discrimination, but are dependent on different mechanisms like learning, generalization, sensory adaptation or synaptic habituation, which may or may not be necessary for a normal odor discrimination.

For instance, the role of nitric oxide signaling was investigated, but its role in olfactory learning (Matsumoto et al, 2006; Muller, 1996) made it difficult to interpret results. However, using an associative learning generalisation paradigm, Hosler, Buxton ans Smith (2000) showed that blockade of NO signaling during learning but not during testing, extends generalization to more dissimilar odors. Using a discriminative learning task in the slug, NO signaling inhibition affects the learning score obtained with similar but not dissimilar odors. Finally, using the masking test, we have seen an effect of NO inhibition on masking effects for certain odor pairs (Colomb, 2006). Different approaches in different model system thus appear to implicate NO in olfactory discrimination, but further analysis is still needed to confirm these results.

In conclusion, the important neurobiological question of how animals discriminate odors is difficult to assess. The three paradigms used so far effectively measure odor similarities, but not directly odor discrimination. Although several studies have begun to investigate the cellular or molecular bases of olfactory discrimination in behaving animals (Doucette et al, 2007; Pavlis et al, 2006; Stopfer et al, 1997), authors have too often disregarded putative effects on learning and memory and their interpretations are sometimes too categorical. The present models of olfactory discrimination mechanisms therefore remain largely hypothetical or based on electrophysiological studies. New studies using different protocols (Cleland et al, 2002) may lead the way toward a better understanding of these mechanisms.

AKNOWLEDGEMENTS

I want to thank Dr. Scaplehorn for his comments on the manuscript and the Swiss National Fund for its funding (grant PBFRA-116951).

REFERENCES

Abarca C, Albrecht U, Spanagel R (2002) Cocaine sensitization and reward are under the influence of circadian genes and rhythm. *Proc Natl Acad Sci U S A* 99(13): 9026-9030.

Borst A (1983) Computation of olfactory signals in *Drosophila melanogaster. J Comp Physiol A* 152: 373-383.

Boyle J, Cobb M (2005) Olfactory coding in Drosophila larvae investigated by cross-adaptation. *J Exp Biol* 208(Pt 18): 3483-3491.

Chandra SB, Singh S (2005) Chemosensory processing in the fruit fly, Drosophila melanogaster: generalization of a feeding response reveals overlapping odour representations. *J Biosci* 30(5): 679-688.

Cleland TA, Morse A, Yue EL, Linster C (2002) Behavioral models of odor similarity. *Behav Neurosci* 116(2): 222-231.

Cobb M (1999) What and how do maggots smell? *Biol Rev*(74): 425-459.

Cobb M, Domain I (2000) Olfactory coding in a simple system: adaptation in Drosophila larvae. *Proc R Soc Lond B Biol Sci* 267(1457): 2119-2125.

Colbert HA, Bargmann CI (1995) Odorant-specific adaptation pathways generate olfactory plasticity. *Neuron* 14(4): 803-812.

Colomb J (2006) The chemosensory system of *Drosophila* larvae: neuroanatomy and behaviour. PhD Thesis, Department of Biology, University of Fribourg, Fribourg (CH).

Colomb J, Grillenzoni N, Stocker RF, Ramaekers A (2007) Complex behavioural changes after odoour exposure in *Drosophila* larvae. *Anim Behav* 73(4): 587-594.

Daly KC, Durtschi ML, Smith BH (2001) Olfactory-based discrimination learning in the moth, Manduca sexta. *J Insect Physiol* 47(4-5): 375-384.

Doucette W, Milder J, Restrepo D (2007) Adrenergic modulation of olfactory bulb circuitry affects odor discrimination. *Learn Mem* 14(8): 539-547.

Faber T, Joerges J, Menzel R (1999) Associative learning modifies neural representations of odors in the insect brain. *Nat Neurosci* 2(1): 74-78..

Fletcher ML, Wilson DA (2002) Experience modifies olfactory acuity: acetylcholine-dependent learning decreases behavioral generalization between similar odorants. *J Neurosci* 22(2): RC201.

Friedrich A, Thomas U, Muller U (2004) Learning at different satiation levels reveals parallel functions for the cAMP-protein kinase A cascade in formation of long-term memory. *J Neurosci* 24(18): 4460-4468.

Guerrieri F, Schubert M, Sandoz JC, Giurfa M (2005) Perceptual and neural olfactory similarity in honeybees. *PLoS Biol* 3(4): e60.

Kadohisa M, Wilson DA (2006) Olfactory cortical adaptation facilitates detection of odors against background. *J Neurophysiol* 95(3): 1888-1896.

Keller A, Vosshall LB (2007) Influence of odorant receptor repertoire on odor perception in humans and fruit flies. *Proc Natl Acad Sci U S A* 104(13): 5614-5619.

Kelliher KR, Ziesmann J, Munger SD, Reed RR, Zufall F (2003) Importance of the CNGA4 channel gene for odor discrimination and adaptation in behaving mice. *Proc Natl Acad Sci U S A* 100(7): 4299-4304.

Korsching S (2002) Olfactory maps and odor images. *Curr Opin Neurobiol* 12(4): 387-392.

L'Etoile ND, Bargmann CI (2000) Olfaction and odor discrimination are mediated by the C. elegans guanylyl cyclase ODR-1. *Neuron* 25(3): 575-586.

Laska M, Galizia CG, Giurfa M, Menzel R (1999a) Olfactory discrimination ability and odor structure-activity relationships in honeybees. *Chem Senses* 24(4): 429-438.

Laska M, Trolp S, Teubner P (1999b) Odor structure-activity relationships compared in human and nonhuman primates. *Behav Neurosci* 113(5): 998-1007.

Lin HH, Lin CY, Chiang AS (2007) Internal representations of smell in the Drosophila brain. *J Biomed Sci.*

Linster C, Johnson BA, Morse A, Yue E, Leon M (2002) Spontaneous versus reinforced olfactory discriminations. *J Neurosci* 22(16): 6842-6845.

Linster C, Johnson BA, Yue E, Morse A, Xu Z, Hingco EE, Choi Y, Choi M, Messiha A, Leon M (2001) Perceptual correlates of neural representations evoked by odorant enantiomers. *J Neurosci* 21(24): 9837-9843.

Matsumoto Y, Unoki S, Aonuma H, Mizunami M (2006) Critical role of nitric oxide-cGMP cascade in the formation of cAMP-dependent long-term memory. *Learn Mem* 13(1): 35-44.

Miyakawa Y (1982) Behavioral evidence for the existence of sugar, salt and amino acid taste receptor cells and some of their properties in Drosophila larvae. *Journal of Insect Physiology* 28(5): 405-410.

Muller U (1996) Inhibition of nitric oxide synthase impairs a distinct form of long-term memory in the honeybee, Apis mellifera. *Neuron* 16(3): 541-549.

Nuttley WM, Atkinson-Leadbeater KP, Van Der Kooy D (2002) Serotonin mediates food-odor associative learning in the nematode Caenorhabditiselegans. *Proc Natl Acad Sci U S A* 99(19): 12449-12454.

Pavlis M, Feretti C, Levy A, Gupta N, Linster C (2006) l-DOPA improves odor discrimination learning in rats. *Physiol Behav* 87(1): 109-113.

Pelz C, Gerber B, Menzel R (1997) Odorant intensity as a determinant for olfactory conditioning in honeybees: roles in discrimination, overshadowing and memory consolidation. *J Exp Biol* 200(Pt 4): 837-847.

Pena T, Pitts RC, Galizio M (2006) Identity matching-to-sample with olfactory stimuli in rats. *J Exp Anal Behav* 85(2): 203-221.

Rodrigues V (1980) Olfactory behavior of Drosophila melanogaster. *Basic Life Sci* 16: 361-371.

Sakura M, Kabetani M, Watanabe S, Kirino Y (2004) Impairment of olfactory discrimination by blockade of nitric oxide activity in the terrestrial slug Limax valentianus. *Neurosci Lett* 370(2-3): 257-261.

Scherer S, Stocker RF, Gerber B (2003) Olfactory learning in individually assayed Drosophila larvae. *Learn Mem* 10(3): 217-225.

Shepard RN (1987) Toward a universal law of generalization for psychological science. *Science* 237(4820): 1317-1323.

Stopfer M, Bhagavan S, Smith BH, Laurent G (1997) Impaired odour discrimination on desynchronization of odour-encoding neural assemblies. *Nature* 390(6655): 70-74.

Vosshall LB, Stocker RF (2007) Molecular architecture of smell and taste in Drosophila. *Annu Rev Neurosci* 30: 505-533.

Wilson DA (2000) Odor specificity of habituation in the rat anterior piriform cortex. *J Neurophysiol* 83(1): 139-145.

Wilson DA, Stevenson RJ (2003) The fundamental role of memory in olfactory perception. *Trends Neurosci* 26(5): 243-247.

Wilson RI, Turner GC, Laurent G (2004) Transformation of olfactory representations in the Drosophila antennal lobe. *Science* 303(5656): 366-370.

Wise PM, Olsson MJ, Cain WS (2000) Quantification of odor quality. *Chem Senses* 25(4): 429-443.

Xia S, Tully T (2007) Segregation of odor identity and intensity during odor discrimination in Drosophila mushroom body. *PLoS Biol* 5(10): 298-307.

Zufall F, Leinders-Zufall T (2000) The cellular and molecular basis of odor adaptation. *Chem Senses* 25(4): 473-481.

In: Animal Behavior: New Research
Editors: E. A. Weber, L. H. Krause

ISBN: 978-1-60456-782-3
© 2008 Nova Science Publishers, Inc.

Chapter 9

CONSTRAINTS AND THE EVOLUTION OF MUTUAL ORNAMENTATION

Ken Kraaijeveld[*] and Barbara M. Reumer

Animal Ecology, Institute of Biology Leiden, Leiden University, PO Box 9516
2300 RA Leiden, The Netherlands

ABSTRACT

Elaborate ornamental traits occur in both males and females of many species. With the accumulation of evidence supporting an adaptive role for female ornaments, the role of genetic and physiological constraints is largely ignored. Here, we investigate phylogenetic patterns of male and female ornamentation in nine disparate animal taxa. In two cases we find evidence consistent with mutual ornamentation resulting from genetic constraints as proposed by Lande (1980), in which sexually monomorphic ornamentation is a temporary stage in the evolution towards sexual dimorphism. However, we also find many cases of mutual ornaments that cannot be explained by this model. Patterns of gains and losses of ornaments in either sex are highly idiosyncratic. We highlight the possibility that some (perhaps many) cases of mutual ornamentation may still be maladaptive, even if they are not the result of genetic correlation as originally envisaged. Physiological constraints could cause maladaptive ornament expression in females as a pleiotropic effect of for example selection on hormone levels. We argue that constraints should be given more attention in empirical studies of mutual ornamentation.

INTRODUCTION

Many animals possess traits that are elaborated beyond what is optimal for survival. Darwin (1871) suggested that such traits can be explained by sexual selection arising through conflicts between males over access to mating opportunities and choosiness on the part of the

---

[*] Email: k.kraaijeveld@biology.leidenuniv.nl

females. Both processes should lead to traits that enhance male competitiveness and/or male attractiveness. When females express these traits, they incur the cost of producing them, but do not gain the benefits. Hence, selection would favour sexually dimorphic expression of such traits. Intense theoretical and empirical research into sexual selection has meant that we now have a well-developed understanding of the evolution of such sexually dimorphic ornamentation in animals (Andersson 1994, Mead & Arnold 2004).

In addition to the well-known examples of sexually dimorphic ornaments, there are many species in which both the male and the female are ornamented. Traditionally, female ornaments were regarded as by-products of sexual selection on males (Muma & Weatherhead 1989). It was thought that because females possess mostly the same genetic material as males, selection for ornaments in males would lead to maladaptive expression of these traits in females. This situation would persist until continued selection resulted in ways to achieve sex-specific expression, which would allow each sex to evolve towards its own optimal expression level (Lande 1980). An alternative route towards sexually dimorphic ornaments, in which genes coding for ornamental traits are sex-linked when they arise, was considered more restrictive and therefore less common (Rice 1984). Thus, cases of sexually monomorphic ornamentation ('mutual ornamentation') were viewed as resulting from sexual selection on males combined with genetic constraints on sex-specific expression.

More recently, the idea that female ornaments may be adaptive has gained popularity. Several studies have shown convincingly that females compete over mating opportunities in some species. Likewise, males can be choosy as to which females they mate with (reviewed in Kraaijeveld et al. 2007). Theoretical work has shown that such mutual sexual selection can indeed result in ornamental traits in both sexes (Servedio & Lande 2006). In addition, social competition in non-sexual contexts may result in selection for elaborate signal traits in both sexes (Kraaijeveld et al. 2007).

The recent flurry of interest in the adaptive function of mutual ornaments has meant a shift in thinking. Many researchers now think that genetic constraints are relatively unimportant. However, the direct evidence for the role of genetic constraints in sex-specific levels of ornament expression is still very limited (Kraaijeveld et al. 2007). The main argument supporting the idea that genetic constraints are unimportant is comparative. Phylogenetic studies in several animal taxa show that female ornaments have been gained and lost relatively frequently, often more so than the homologous traits in males (Amundsen 2000, Wiens 2001). This has been taken as evidence that female ornaments are relatively unconstrained genetically.

Here, we study phylogenetic patterns of male and female ornamentation in a wide variety of animal groups. First, we test the prediction by Lande (1980) that mutual ornamentation is a temporary stage in the evolution towards sexual dimorphism. Second, we explore the argument that frequent gain and loss of female ornaments should be interpreted as lack of genetic constraints.

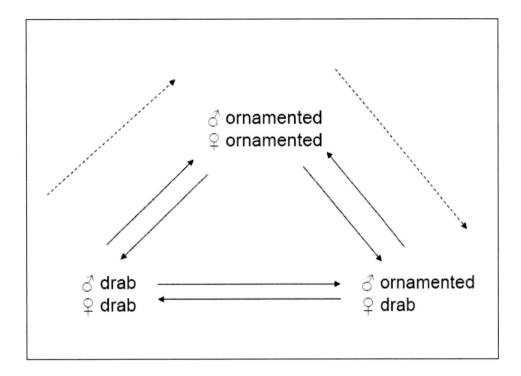

Figure 1. Schematic representation of the six possible transitions between character state combinations. The broken lines show the sequence of transitions predicted by the model of Lande (1980).

METHODS

When treated as dichotomous, ornamental traits can take three states with respect to sexual dimorphism: male and female both drab, male and female both ornamented, male ornamented, female drab. We ignore here the much rarer situation where the female is ornamented, but the male drab. On a phylogeny in which all three character states are present, there are six possible transitions between states (Figure 1). To study the relative frequency and phylogenetic position of these transitions, we searched the literature for phylogenetic trees containing all three character states. Our search was not exhaustive, but rather aimed at identifying informative groups in a wide range of animal taxa. Character states were mapped onto the phylogeny using MacClade (Maddison & Madison 2001).

RESULTS

Stalk-eyed Flies

Stalk-eyed flies of the family Diopsidae are characterized by having their eyes at the ends of stalks. Male eye stalk length is involved in male-male competition over access to females (Panhuis & Wilkinson 1999) and females prefer to mate with males with long eye stalks

(Hingle et al. 2001). The only study to investigate whether female eye stalks in a sexually dimorphic species is adaptive, found negative results (Al-khairulla et al. 2003). The length of the eye stalks and the degree of sexual dimorphism vary considerably between species. Baker & Wilkinson (2001) mapped the degree of sexual dimorphism onto a molecular phylogeny of 30 species. Baker & Wilkinson (2001) quantified the degree of sexual dimorphism by comparing the eye span-body length regression slopes of males and females. A consequence of this method is that species in which the male and female have very different eye span will be classified as monomorphic as long as the allometric slopes are the same. Furthermore, we needed to distinguish between monomorphic ornamented and monomorphic non-ornamented species for our analysis. We therefore re-classified the species by calculating the difference in eyespan-body length ratio between males and females. Species with a ratio below 0.25 were classified as non-ornamented, those with higher ratios as ornamented. A species was classified as sexually monomorphic if the eyespan-body length ratio differed between males and females by less than 0.1 and dimorphic if the ratio exceeded 0.1.

Elongated eye stalks evolved from normal heads at the base of the phylogenetic tree (Figure 2). The species in the most basal group with elongated eye stalks (the genus *Sphyracephala*) are mostly sexually monomorphic or only weakly dimorphic. Extreme levels of sexual dimorphism are more derived. The basal evolution towards highly elaborate, sexually dimorphic eye span thus follows the sequence predicted by Lande (1980).

While the sexual monomorphism of the *Sphyracephala* clade and other basal diopsids may be viewed as a primitive state on the evolutionary pathway towards increased dimorphism, at least two other monomorphic species appear at higher nodes (Figure 2). As these species have dimorphic ancestors, the genetic architecture for dimorphism was presumably present when these species evolved.

Dung Beetles

Dung beetles of the genus *Onthophagus* display a variety of horn-like outgrows on the head and thorax. Species vary in the extent of sexual dimorphism in these traits. Male horns are used in combat over access to mates. Females occasionally fight with other females over the ownership of breeding tunnels, but to what extent this selects for horns in females has not been investigated. Emlen et al. (2005) mapped five types of horns onto a molecular phylogeny of 31 species of *Onthophagus*. We assume that each horn type is homologous between species and treat each type separately, focussing on horns at the base of the head.

Horns at the base of the head are ancestral to this genus, precluding investigation of the early evolution of this horn type (Figure 3). However, three other types of horn appear not to be ancestral: those on the center and front of the head and on the center of the thorax (Emlen et al. 2005). Two of these three traits first appear in sexually dimorphic state. Only the horns on the center of the head are expressed in both males and females when they first appear, albeit not to the same extent. Thus, genes for these traits were either male-limited at the outset, or were co-opted by modifiers relatively quickly.

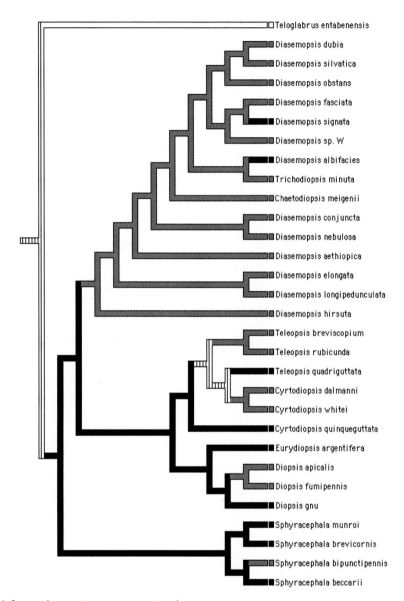

☐ Male and female not ornamented
▨ Male ornamented, female not ornamented
■ Male and female ornamented
▥ Equivocal

Figure 2. Ornamentation (relative eyespan) mapped onto the phylogeny of stalk-eyed flies. After Baker & Wilkinson (2001).

Male and female not ornamented

Male ornamented, female not ornamented

Male and female ornamented

Equivocal

Figure 3. Ornamentation (horns at the base of the head) mapped onto the phylogeny of dung beetles of the genus *Onthophagus*. After Emlen et al. (2005).

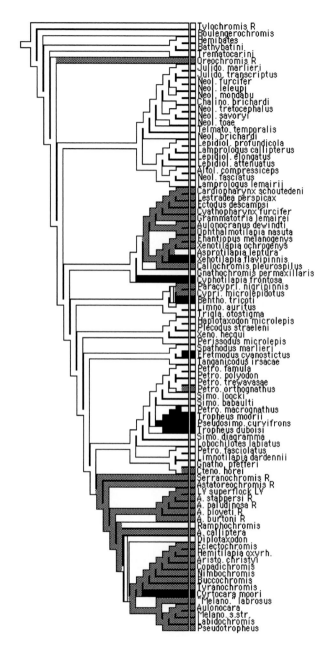

☐ Male and female not ornamented
▦ Male ornamented, female not ornamented
■ Male and female ornamented
▥ Equivocal

Figure 4. Ornamentation (nonmelanic colouration) mapped onto the phylogeny of African cichlid fish. After Seehausen et al. (1999).

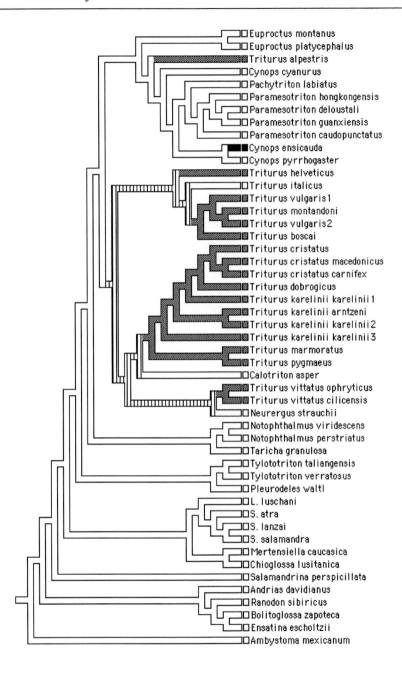

Male and female not ornamented

Male ornamented, female not ornamented

Male and female ornamented

Equivocal

Figure 5. Ornamentation (dorsal crests) mapped onto the phylogeny of newts of the genus *Triturus* and related genera. Phylogeny after Steinfartz et al (2006).

When the entire phylogeny was considered, male horns at the base of the head were lost at least five times and regained twice (Figure 3). Female horns were gained three times from sexually dimorphic ancestors, each time resulting in mutual ornamentation.

Cichlid Fish

African cichlid fish (Cichlidae) are a very species-rich group that show wide variety in colour patterns in both sexes. Non-melanic colours in males have been linked to female mate preference in some species (Maan et al. 2004), but little is known about the function (if any) of such colours in females. Seehausen et al. (1999) constructed a composite phylogeny for 89 species of African cichlids and mapped changes in sexual dimorphism in non-melanic colours. We inspected photos on the internet to assess whether the sexually monomorphic species where dull or colourful.

Both sexes are dull in the most basal groups in the phylogeny (Figure 4). This is followed by a eight or nine gains of colourfulness. In at least four of these gains, only the male is colourful (i.e. it appears as a sexually dimorphic trait), while in at least three cases colourfulness appears in males as well as females. In at least one additional clade, the ancestral state is equivocal. Because colourfulness appears in sexually dimorphic state deep in the phylogeny, before the first appearance of sexually monomorphic colourfulness, the patterns in cichlids do not unequivocally support the Lande process.

All possible transitions between character states are present in this phylogeny (Figure 4). Male colourfulness is lost secondarily from dimorphic ancestors three times, resulting in monomorphic drab species. Female colourfulness is gained from dimorphic ancestors at least twice, leading to mutually ornamented species.

Newts

Newts of the genus *Triturus* and their close relatives possess dorsal combs ('crests') that are very elaborate in the males of certain species. Experiments have shown that larger crests are preferred by females in one species (Malacarne & Cortassa 1983). We mapped male and female dorsal crests onto the molecular phylogeny of Steinfartz et al. (2006). Dorsal crests were scored as present or absent based on photos on the internet.

Based on this phylogeny, dorsal crests appear at least twice, in both cases as a sexually dimorphic trait (Figure 5). Crests are subsequently lost several times (Figure 5). The only species in which both sexes display a small dorsal crest occurs in a clade of otherwise uncrested Asian newts and is thus not ancestral to the crested clades. The lack of sexually monomorphic species at the base of the crested clades means that this group of newts does not show signs of the pattern predicted from the Lande model.

Female crests are very rare and there are no transitions between dimorphic and monomorphic ornamented states or vice versa.

Lizards

Males of many species of phrynosomatid lizard have conspicuously coloured (usually blue) patches on the sides of the belly and throat. Colourful belly patches also occur in females of some species, and are absent from both sexes in others. The belly patches are important in male-male competition and have been implicated in female mate preferences (Cooper & Greenberg 1992). Whether female belly patches are also functional is unknown. Wiens (1999) mapped these character states onto a phylogenetic tree of 130 taxa.

Inspection of the basal nodes of the phylogenetic tree reveals that belly patches may have evolved either once or twice from a monomorphic dull state (Figure 6). In each case, the appearance of sexually dimorphic patches is preceded by belly patches in both sexes (in *Uma* and *Uta*), consistent with the Lande model.

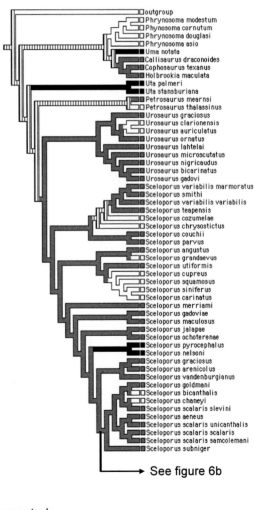

See figure 6b

☐ Male and female not ornamented
▨ Male ornamented, female not ornamented
■ Male and female ornamented
▥ Equivocal

a)

Figure 6 (Continued on next page.)

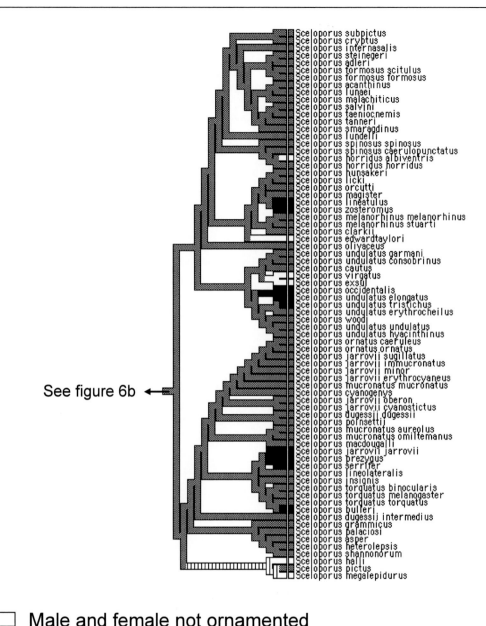

See figure 6b ◄—

☐ Male and female not ornamented
▨ Male ornamented, female not ornamented
■ Male and female ornamented
▥ Equivocal

b)

Figure 6. Ornamentation (coloured belly patches) mapped onto the phylogeny of phrynosomatid lizards. After Wiens (1999).

Sexual dimorphism was lost secondarily at least 16 times. In six cases this occurred through the gain of female belly patches (resulting in monomorphic ornamentation), while in ten cases the males lost the belly patch (resulting in monomorphic drab species).

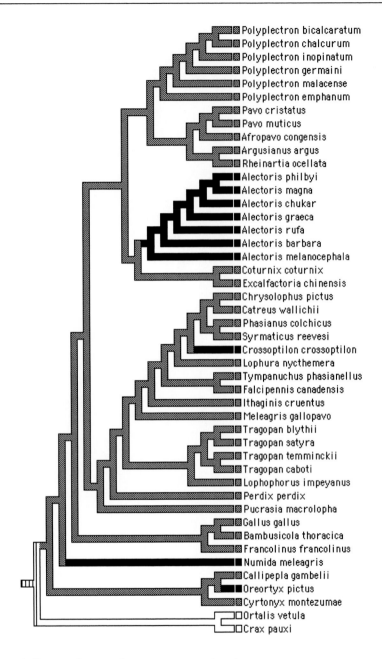

☐ Male and female not ornamented
▨ Male ornamented, female not ornamented
■ Male and female ornamented
▥ Equivocal

Figure 7. Ornamentation (plumage colour) mapped onto the phylogeny of pheasants and relatives. Phylogeny after Kimball et al. (2001).

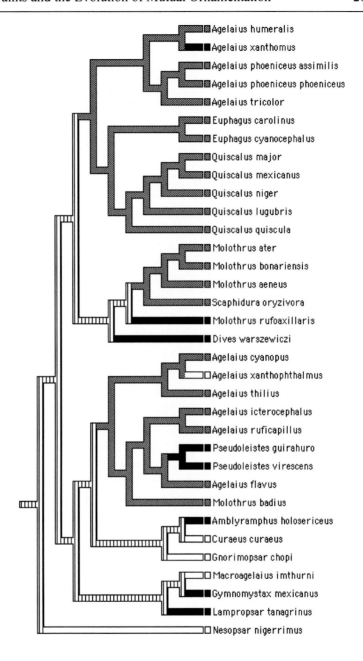

Figure 8. Ornamentation (plumage colour) mapped onto the phylogeny of New World blackbirds. Phylogeny after Lanyon & Omland (1999).

Pheasants

Galliform birds, including pheasants and peafowl, provide some of the most striking examples of male ornamentation among animals. Females of some groups are also elaborately coloured. Male plumage elaboration is important in female mate choice and male-male competition in several species (Mateos & Carranza 1997, Hagelin 2002, Loyau et al. 2005), but female plumage colour has not been studied. For 46 species of Galliformes, we scored male and female plumage colour as being colourful or not, based on photos on the internet. These character states were then mapped onto the phylogeny by Kimball et al. (2001).

Colourful plumage probably evolved from a monomorphically drab ancestor and first appeared as a sexually dimorphic trait (Figure 7). Sexually monomorphic ornamentation evolved four times, each through the gain of female ornamentation from a dimorphic ancestor. Thus the pattern among Galliformes does not conform to the Lande model.

Icterids

New World blackbirds (Icterinae) display conspicuous plumage colour that may be present in both sexes, males only, or neither. Colourful plumage patches in one well-studied member of this group, the Red-winged blackbird *Agelaius phoeniceus*, have been shown to be involved in male social dominance (Searcy 1979). We classified male and female plumage of 34 icterid taxa as colourful or drab based on information in Irwin (1994) and photos on the internet. The scores were mapped onto the phylogeny of Lanyon & Omland (1999).

The most basal taxon in the phylogeny is monomorphic and dull (Figure 8). The basal taxa in at least two of the four higher subgroups are monomorphic and colourful. Sexual dimorphism is a derived trait. The basic pattern is thus partially consistent with the Lande model. Nevertheless, from the sexually dimorphic state, female colourfulness is gained twice and male colourfulness lost once.

Tanagers

Like the Icterids, the tanagers represent a large clade of New World birds with variable male and female ornamentation. Burns (1998) scored 47 species of tanager for plumage brightness and sexual dimorphism and mapped these onto a phylogeny. Six of the species were scored as polymorphic and were omitted from the present analysis.

The state of sexual dimorphic ornamentation appears to be ancestral in this group (Figure 9). With drabness not being ancestral, Lande's model cannot be traced on the phylogeny. Monomorphic states (drab and colourful) have evolved repeatedly from various ancestral states. However, sexual dimorphism was not regained once lost.

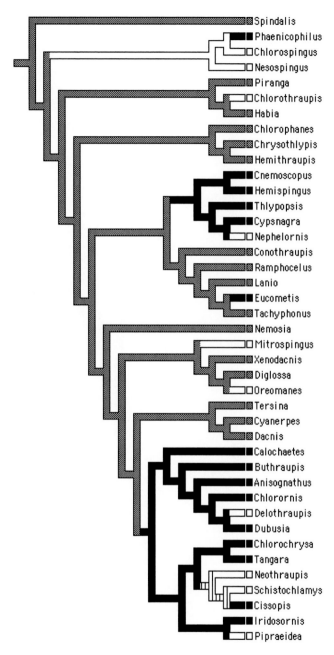

Male and female not ornamented
Male ornamented, female not ornamented
Male and female ornamented
Equivocal

Figure 9. Ornamentation (plumage colour) mapped onto the phylogeny of tanagers. After Burns (1998).

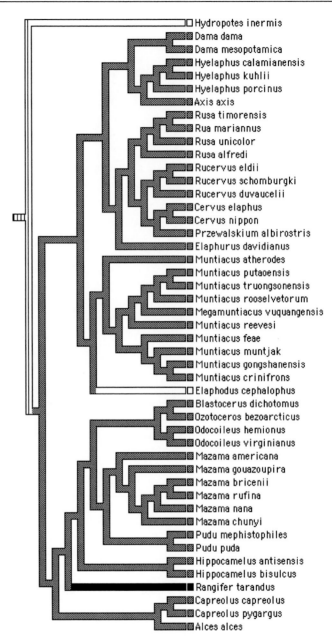

☐ Male and female not ornamented

▦ Male ornamented, female not ornamented

■ Male and female ornamented

▥ Equivocal

Figure 10. Ornamentation (antlers) mapped onto the phylogeny of deer. Phylogeny after Fernández & Vrba (2005).

Deer

The antlers of male deer (Cervidae) are an iconic example of display traits resulting from sexual selection through male-male combat. Furthermore, Ditchkoff et al. (2001) showed that antler development is an honest signal of individual genetic quality, suggesting it could be used in female mate choice. Fernández & Vrba (2005) provide a composite phylogeny for the Cervidae, which we used to map the presence/absence of antlers in males and females. To assess whether the species had antlers or not, we inspected the plates in Grzimeck (1975).

The ancestor to the deer probably had no antlers and almost all derived taxa are sexually dimorphic (Figure 10). Female antlers occur in reindeer (*Rangifer tarandus*), a derived taxon, and neither sex has antlers in Tufted deer (*Elaphodus cephalophus*), another derived taxon. Thus, the lack of sexually monomorphic taxa among the basal cervids means that Lande's model cannot be traced on the phylogeny of this group.

CONCLUSION

Lande's (1980) Model and Genetic Constraints on Female Ornamentation

Lande's (1980) model in which mutual ornamentation is presented as a temporary stage in the evolution towards sexual dimorphism is often cited as a null hypothesis in the study of monomorphic ornaments; the alternative being an adaptive function for both the male and the female ornament (Amundsen 2000). Lande's (1980) model describes the evolution of a novel (ornamental) trait. As such, it makes the prediction that a truly novel ornament should arise in a sexually monomorphic state. If the subsequent evolution of sexual dimorphism in the ornamental trait is slow (relative to species formation), it might be possible to trace such a pattern on phylogenies of extant species. Two of the phylogenies investigated in this paper show patterns that are consistent with Lande's (1980) model. These are the stalk-eyed flies and the lizards. Patterns displayed by the Icteridae were less clear-cut, but may also be partially consistent with the model. In both cases the appearance of sexual dimorphism is preceded by sexually monomorphic ornamentation. Thus, sexually monomorphic ornaments in the stalk-eyed fly genus *Sphyracephala* and the lizard genuses *Uma* and *Uta* may represent a primitive state in which the female ornament is the result of a genetic correlation with the male ornament.

Our failure to visualize Lande's (1980) prediction on some of the phylogenies we investigated does not mean this model does not adequately represent the sequence of events in these taxa. It may simply mean that the basal process that the model describes has been erased by subsequent evolution. For example, antlers may have appeared as a sexually monomorphic state in the basal Cervids, but all members of such a clade may have become extinct. However, in some cases the ornament seems to have been sexually dimorphic from the start. The most convincing example is the appearance of horns *on the front of the head* in the *Onthophagus* beetles (note that this is not the horn type represented in Figure 3). This trait only appears in two species on the outer branches of the phylogeny from ancestors in which neither sex has this horn type, suggesting it is a recent novelty. Both species are sexually dimorphic for the trait, suggesting that the trait appeared as such. It would be interesting to conduct a wider survey of sexual dimorphism in such recent ornamental novelties.

In none of the phylogenies investigated do the patterns of evolutionary change conform to the Lande (1980) model when all changes are taken into account (not just those at the base of the tree). Instead, overall rates of change between character states are highly idiosyncratic. Our results emphasize an important point, namely that Lande's (1980) model does *not* provide a general explanation for cases of mutual ornaments among extant taxa. This is so because the direct ancestors of many mutually ornamented taxa were sexually dimorphic for the homologous trait. In such cases (which are frequent) the ornament became sexually monomorphic while the genetic architecture for sexual dimorphism was present.

Threshold-mechanisms and Sex-specific Ornament Expression

Our conclusion that Lande's (1980) model does not provide a general explanation for extant cases of mutual ornamentation does *not* mean that constraints are unimportant in the evolution of female ornaments. Instead, the patterns we find are consistent with threshold mechanisms for ornament expression (West-Eberhard 1989, 2003, Emlen et al. 2005). In such mechanisms, the level of gene expression is coupled to environmental cues, mediated through hormone levels (see Emlen et al. 2005 and references therein). Sexual dimorphism may then result from different levels of circulating hormones in males and females.

Threshold mechanisms suggest a role for physiological constraints in the evolution of mutual ornaments (Emlen et al. 2005). Suppose a system where ornament expression is dependent on circulating levels of a hormone, say testosterone (T). Below a certain threshold level of T, the ornament is not expressed. T levels in females are below the threshold, in males above it. Imagine now that selection on something other than the ornamental trait causes the hormonal balance in females to shift. For example, selection for increased aggression in females may promote increased levels of T (or increased sensitivity to T) in females. If the level of T in females is raised beyond the threshold of ornament expression, the ornament will be expressed as a pleiotropic effect of selection on aggression. Such pleiotropic expression is then the result of a physiological constraint and may be maladaptive. Further selection to raise the threshold in females will then reinstate sexual dimorphism, as in the second phase of the Lande (1980) process. The position of mutually ornamented species at the tips of phylogenetic branches in a number of the examples we studied (e.g. the lizards) supports this idea. It seems likely that in these cases not enough time has passed to allow selection to alter the threshold level in females. A similar argument can be made for the loss of the male ornament through hormone levels in males dropping below the threshold for ornament expression.

Although superficially similar, the threshold mechanism predicts a different sequence of character states than the Lande process (Figure 11). The frequent transitions in both directions between sexually dimorphic and monomorphic (dull and bright) states as seen in several of the phylogenies (stalk-eyed flies, beetles, lizards) are consistent with developmental thresholds. We argue that the threshold mechanism provides a realistic non-adaptive explanation for the evolution of many cases of mutual ornamentation.

Lande (1980) process

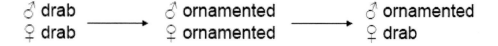

Treshold mechanism

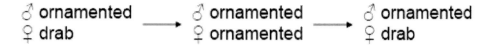

Figure 11. Sequence of character state transitions predicted by the Lande (1980) and threshold models.

Lande's (1980) model and physiological constraints are not mutually exclusive processes. As we have argued, Lande's model deals with novel ornaments, so should only be present when the trait first appears (relatively deep nodes in most of the phylogenies that we discussed here). On the other hand, maladaptive female ornaments due to threshold mechanisms are expected at the tips of phylogenies, because they should be short-lived. Thus both mechanisms may potentially be present in a single phylogeny. Those of the stalk-eyed flies and the lizards discussed in this chapter are possible examples.

Many cases of mutual ornamentation are likely to be adaptive (reviewed in Kraaijeveld et al. 2007). However, this study shows that the role of genetic and physiological constraints should not be disregarded. Detailed estimates of genetic correlations and their role in constraining adaptive ornament expression are sorely needed. The present study shows that investigation of the phylogenetic position of the study organism may provide insights into the relative likelihood of various types of constraints.

REFERENCES

Al-khairulla, H., Warburton, D. & Knell, R. J. (2003). Do the eyestalks of female diopsid flies have a function in intrasexual aggressive encounters? *Journal of Insect Behavior,* 16, 679-686.

Amundsen, T. (2000). Why are female birds ornamented? *Trends in Ecology and Evolution,* 15, 149-155.

Andersson, M. B. (1994). *Sexual selection.* Princeton, Princeton University Press.

Baker, R. H. & Wilkinson, G. S. (2001). Phylogenetic analysis of sexual dimorphism and eye-span allometry in stalk-eyed flies (Diopsidae). *Evolution,* 55, 1373-1385.

Burns, K. J. (1998). A phylogenetic perspective on the evolution of sexual dichromatism in tanagers (Thraupidae), the role of female versus male plumage. *Evolution,* 52, 1219-1224.

Cooper, Jr W. E. & Greenberg, N. (1992). Reptilian coloration and behavior. In, *Biology of the reptilian. 18. Hormones, brain, and behavior.* (eds C. Gans & D. Crews), pp. 298-422. Chicago, University of Chicago Press.

Darwin, C. (1871). *The descent of man; and selection in relation to sex.* New York, Crowell..

Ditchkoff, S. S., Lochmiller, R. L., Masters, R. E., Hoofer, S. R. & Van Den Bussche, R. A. (2001). Major-histocompatibility-complex-associated variation in secondary sexual traits of white-tailed deer (*Odocoileus virginianus*), evidence for good-genes advertisement. *Evolution,* 55, 616-625.

Emlen, D. J., Hunt, J. & Simmons, L. W. (2005). Evolution of sexual dimorphism and male dimorphism in the expression of beetle horns, phylogenetic evidence for modularity, evolutionary lability, and constraint. *American Naturalist,* 166, S42-S67.

Fernández, M. H. & Vrba, E. S. (2005). A complete estimate of the phylogenetic relationships in Ruminantia, a dated species-level supertree of the extant ruminants. *Biological Reviews,* 80, 269-302.

Grzimeck, B. (1975). *Grzimeck's Animal Life Encyclopedia,* Volume 11, Mammals II. New York, Van Nostrand Reinhold Company.

Hagelin, J. C. (2002). The kinds of traits involved in male-male competition, a comparison of plumage, behavior, and body size in quail. *Behavioural Ecology,* 13, 32-41.

Hingle, A., Fowler, K. & Pomiankowski, A. (2001). Size-dependent female mate preference in the stalk-eyed fly, *Cyrtodiopsis dalmanni. Animal Behaviour,* 61, 589-595.

Irwin, R. A. (1994). The evolution of plumage dichromatism in the New World blackbirds, social selection on female brightness? *American Naturalist,* 144, 890-907.

Kimball, R. T., Braun, E. L., Ligon, J. D., Lucchini, V. & Randi, E. (2001). A molecular phylogeny of the peacock-pheasants (Galliformes, *Polyplectron* spp.) indicates loss and reduction of ornamental traits and display behaviours. *Biological Journal of the Linnean Society,* 73, 187-198.

Kraaijeveld, K., Kraaijeveld-Smit, F. J. L. & Komdeur, J. (2007). The evolution of mutual ornamentation. *Animal Behaviour,* 74, 657-677.

Lande, R. (1980). Sexual dimorphism, sexual selection, and adaptation in polygenic characters. *Evolution,* 34, 292-305.

Lanyon, S. M. & Omland, K. E. (1999). A molecular phylogeny of the blackbirds (Icteridae), five lineages revealed by cytochrome-B sequence data. *Auk,* 116, 629-639.

Loyau, A., Jalme, M. S. & Sorci, G. (2005). Intra- and intersexual selection for multiple traits in the peacock (*Pavo cristatus*). *Ethology,* 111, 810-820.

Maan, M., Seehausen, O., Söderberg, L., Johnson, L., Ripmeester, E. A. P., Mrosso, H. D. J., Taylor, M. I., Van Dooren, T. J. M. & van Alphen, J. J. M. (2004). Intraspecific sexual selection on a speciation trait, male coloration, in the Lake Victoria cichlid *Pundamilia nyererei. Proceedings of the Royal Society of London* B, 271, 2445-2452.

Maddison, D. R. & Maddison, W. P. (2001). *MacClade 4, analysis of phylogeny and character evolution.* Version 4.03. Sinauer Associates, Sunderland, Massachusetts.

Malacarne, G. & Cortassa, R. (1983). Sexual selection in the crested newt. Animal Behaviour, 31, 1256-1264.

Mateos, C. & Carranza, J. (1997). The role of bright plumage in male-male interactions in the ring-necked pheasant. *Animal Behaviour,* 54, 1205-1214.

Mead, L. S. & Arnold, S. J. (2004). Quantitative genetic models of sexual selection. *Trends in Ecology and Evolution,* 19, 264-271.

Muma, K. E. & Weatherhead, P. J. (1989). Male traits expressed in females, direct or indirect selection? *Behavioral Ecology and Sociobiology,* 25, 23-31.

Panhuis, T. M. & Wilkinson, G. S. (1999). Exaggerated eye span influences male contest outcome in stalk-eyed flies. *Behavioral Ecology and Sociobiology,* 46, 221-227.

Rice, W. R. (1984). Sex chromosomes and the evolution of sexual dimorphism. *Evolution,* 38, 735-742.

Searcy, W. A. (1979). Morphological correlates of dominance in captive male Red-winged blackbirds. *Condor,* 81, 417-420.

Seehausen, O., Mayhew, P. J. & van Alphen, J. J. M. (1999). Evolution of colour patterns in East African cichlid fish. *Journal of Evolutionary Biology,* 12, 514-534.

Servedio, M. R. & Lande, R. (2006). Population genetic models of male and mutual mate choice. *Evolution,* 60, 674-685.

Steinfartz, S., Vicario, S., Arntzen, J. W. & Caccone, A. (2006). A Bayesian approach on molecules and behavior, reconsidering phylogenetic and evolutionary patterns of the Salamandridae with emphasis on *Triturus* newts. *Journal of Experimental Zoology,* 308B, 139-1662.

West-Eberhard, M. J. (1989). Phenotypic plasticity and the origins of diversity. *Annual Reviews of Ecology and Systematics,* 20, 249-278.

West-Eberhard, M. J. (2003). *Developmental plasticity and evolution.* Oxford, Oxford University Press.

Wiens, J. J. (1999). Phylogenetic evidence for multiple losses of a sexually selected character in phrynosomatid lizards. *Proceedings of the Royal Society of London B,* 266, 1529-1535.

Wiens, J. J. (2001). Widespread loss of sexually selected traits, how the peacock lost its spots. *Trends in Ecology and Evolution,* 16, 517-523.

INDEX

A

abdomen, 95, 96, 123
abiotic, 149
access, 3, 17, 36, 39, 119, 121, 122, 136, 141, 142, 143, 145, 146, 147, 148, 170, 171, 172, 173, 175, 193, 195, 196
accessibility, 171
accuracy, 187
acetylcholine, 191
acid, 191
acoustic signals, ix, 93
acoustical, 155
activation, 62, 65, 66, 68, 69, 70, 71
adaptation, ix, x, 27, 93, 95, 121, 124, 125, 127, 129, 185, 186, 187, 188, 189, 190, 191, 192, 212
addiction, 63, 71
adjustment, 119, 131
administration, 65
adrenoceptors, 70
adult(s), 19, 21, 22, 23, 38, 46, 55, 94, 97, 98, 99, 103, 105, 106, 107, 118, 119, 122, 125, 135, 140, 143, 177
adulthood, ix, 2, 93, 98, 99
advertising, 28
Africa, 182
age, 41, 48, 50, 52, 54, 66, 74, 76, 78, 79, 81, 126, 136, 152, 154, 179
aggression, 20, 29, 58, 59, 67, 68, 69, 70, 170, 173, 210
aggressive behavior, 59
aging, 53
aid, 144
aircraft, 147
Alaska, 152, 153, 156, 157, 159
alleles, 43, 44
allometry, 211

alternative(s), ix, 74, 93, 95, 108, 120, 122, 123, 124, 130, 143, 144, 154, 156, 160, 170, 194, 209
ambivalence, 67
amino acid, 191
amphibia, 165
amplitude, 140, 144
anabolic, 69
androgens, 71
animal communication, vii
animal models, 186
animals, vii, ix, x, 1, 46, 47, 58, 59, 67, 68, 74, 75, 86, 87, 89, 91, 133, 145, 146, 165, 167, 168, 171, 172, 173, 174, 183, 185, 186, 187, 188, 190, 193, 206
ANOVA, 103
antagonistic, 49
Antarctic, ix, x, 133, 134, 138, 147, 148, 152, 155, 157, 158, 159, 177
anthropogenic, 145, 151
antidepressants, 69
anti-predator, ix, 74, 87
ants, 129
anxiety, 58, 64, 66, 69, 70
aquatic, 134, 141, 160
Arctic, ix, 133, 135, 137, 138, 145, 157, 158, 159, 160
Arctic Ocean, 135
argument, 45, 194, 210
arousal, 58, 59, 62, 63, 64, 68, 69, 71
arthropods, 94, 127, 174
Asia, 145
Asian, 201
assessment, 12, 25, 47
assumptions, 9, 11, 24, 49
asymmetry, 3, 4, 12, 29, 30, 32, 189
attacks, 63, 96, 113
attention, viii, xi, 5, 17, 24, 46, 47, 73, 74, 78, 87, 90, 91, 95, 193

attractiveness, 10, 12, 19, 22, 29, 30, 36, 37, 45, 121, 194
audio, 145, 149
Australia, 113, 114, 115, 125, 164
Austria, 73
Autonomous, 139, 146, 161
availability, 118, 121, 136, 142, 144, 151, 168, 170, 174
aversion, 20
avoidance, 19, 20, 29, 30, 39, 42, 44, 46, 50, 51, 67, 88, 91
avoidance behavior, 67
awareness, 63

B

background noise, 151
Baikal, 135, 138, 140, 144
bandwidth, 146, 147, 157
barriers, viii, 73, 74, 75, 76, 77, 78, 80, 83, 84, 88, 89, 91
basic research, 150
battery(ies), 146, 148
Bayesian, 213
beetles, 196, 198, 209, 210
behavior, vii, viii, 33, 48, 55, 57, 58, 59, 60, 61, 62, 63, 64, 65, 66, 67, 68, 69, 70, 71, 125, 128, 130, 153, 155, 156, 158, 188, 192, 212, 213
benefits, vii, 1, 5, 7, 10, 13, 14, 15, 16, 17, 19, 22, 35, 36, 37, 38, 39, 40, 44, 45, 46, 49, 53, 54, 56, 170, 180, 194
benzodiazepine, 63
bias, 2, 22, 44, 75
bilateral, 22
biogeography, 96, 128, 183
biological, xi, 22, 53, 68, 94, 124, 130, 185
biological control, 94, 124, 130
biology, vii, 1, 124, 125, 127, 140, 177, 178, 182, 193
biomechanics, 130
biota, 177
birds, viii, x, 11, 16, 28, 37, 40, 41, 73, 74, 76, 78, 79, 80, 81, 82, 83, 84, 85, 86, 87, 88, 90, 126, 163, 164, 165, 167, 168, 170, 171, 172, 174, 175, 176, 177, 178, 179, 181, 182, 183, 206, 211
birth, vii, 4, 29, 35, 40, 135, 140, 141, 142, 144, 145
bison, 39, 56
blends, 186, 189
blocks, 75, 79
blood, viii, 11, 57, 59, 60, 62, 65
body size, ix, 43, 46, 47, 49, 93, 94, 95, 96, 97, 117, 121, 122, 123, 129, 131, 134, 165, 212
body weight, 29, 101, 103, 123

bonds, 165
Boston, 91, 154, 158
brain, x, 63, 65, 68, 69, 71, 77, 185, 186, 191, 212
brain size, 77
brain structure, 63
Brazil, 111, 113
breast, 10, 164
breathing, 95, 97, 99, 118, 124
breeding, vii, ix, x, 5, 7, 8, 18, 35, 39, 47, 49, 53, 96, 113, 133, 134, 135, 136, 137, 140, 141, 142, 143, 144, 145, 151, 152, 154, 155, 156, 157, 158, 161, 163, 164, 165, 168, 170, 173, 175, 178, 183, 196
Britain, 153
broadband, 142, 143, 148
burns, 163
bushcricket, ix, 93, 94, 95, 96, 97, 112, 113, 116, 118, 119, 121, 122, 123, 124, 125, 128, 129
by-products, 194

C

C. frugilegus, viii, 73
C. monedula, viii, 73
cables, 148
cache, x, 76, 77, 88, 89, 163, 166, 167, 168, 170, 171, 172, 173, 174, 175, 176, 179, 181, 182, 183
California, 154, 158
cAMP-protein kinase A, 191
Canada, 153, 156, 159, 160
canopies, x, 163, 167
capacity, 75, 121, 139, 146, 147, 149
carbon, 186, 189
Caspian, 135, 138, 140, 144
cast, viii, 57
catharsis, 62
catholic, 174
cats, 165
causation, 51
cave, 182
certainty, 38, 60
channels, 149
chemical, 117, 187, 189
chemotaxis, xi, 185, 186
Chicago, 27, 32, 49, 55, 178, 212
chickadees, x, 89, 91, 163, 169, 174, 178, 181, 183
chickens, 90
childhood, 20
children, 74
chimpanzee(s), viii, 20, 27, 33, 39, 50, 52, 55, 56, 73, 75, 87, 91, 92
chromosomes, 213
chronic stress, 66, 69
circadian, 190

clams, 175
classification, x, 134, 149, 154
cleaning, 78
climate change, 136, 144, 160
clusters, 142
coal, 180, 182
cockroaches, 53
coding, x, 185, 190, 194
coercion, 32, 39, 40, 45
cognition, 58, 180
cognitive, 74, 76, 77, 87, 90, 175, 176, 180
cognitive ability(ies), 74, 176
cognitive development, 90
cognitive research, 74
cognitive science, 74
cognitive tasks, 175
cognitive tool, 176
collisions, 149
colonisation, 165
combat, 196, 209
combined effect, 47
commercial, 145
common rule, 67
communication, vii, ix, 74, 117, 122, 124, 127, 133, 136, 141, 142, 143, 145, 147, 157, 177
communication systems, 127
community(ies), 50, 94, 153
comparative research, 77, 89
compatibility, 43, 53
competition, ix, 2, 3, 4, 5, 17, 25, 27, 31, 36, 37, 39, 40, 43, 44, 45, 48, 49, 52, 55, 56, 71, 93, 94, 119, 120, 122, 123, 125, 130, 131, 141, 143, 171, 194, 195, 202, 206, 212
competitive advantage, 2
competitiveness, 194
competitor, 171
complement, 43
complex interactions, 136
complexity, 11, 65, 151, 156
components, 41, 47, 54, 60, 61, 62, 63, 64, 66, 67, 68
composite, 54, 201, 209
composition, 177
computer technology, 145
concentration, xi, 185, 186, 187, 189
conception, 4, 39, 46, 63, 181
concordance, 79
conditioning, 20, 192
confidence interval, 85
configuration, 149
conflict, 4, 56, 170, 179
confusion, 140
conjecture, 37, 176
conscious awareness, 63

conservation, 166, 168, 178
consolidation, 192
constraints, xi, 2, 36, 40, 46, 54, 193, 194, 210, 211
consumption, 63, 156, 167, 170, 180
continuity, 88
control, viii, 11, 35, 37, 40, 41, 46, 59, 60, 65, 66, 67, 68, 71, 78, 79, 81, 82, 83, 84, 85, 94, 124, 130, 170
control group, 11
convergence, 130
copulation, 14, 15, 18, 39, 44, 45, 52, 54, 63, 64
correction factors, 146
correlation(s), viii, xi, 5, 7, 8, 10, 11, 13, 26, 31, 39, 40, 57, 59, 60, 61, 64, 115, 119, 149, 156, 193, 209, 211
cortex, 192
cortical, 70, 191
cortisol, 71, 152
corvids, viii, x, 73, 77, 87, 88, 89, 90, 163, 172, 175, 183
Corvus corax, viii, 73, 90, 91
costs, 6, 9, 14, 20, 22, 37, 38, 39, 42, 43, 46, 148, 151, 170, 171
coupling, 130
coverage, 146, 147
covering, 147
crack, 147
craving, 66, 71
crows, x, 88, 91, 163, 175, 183
cues, viii, 41, 42, 43, 46, 47, 48, 49, 54, 71, 73, 76, 78, 81, 82, 83, 84, 85, 86, 87, 91, 95, 117, 121, 125, 136, 137, 140, 155, 160, 210
culture, vii
cycles, 168, 170
cytochrome, 212

D

data analysis, 147
data collection, 147
data set, x, 133, 134, 145
data transfer, 148
decay, 167, 174, 175
deception, 77
deciduous, 181
decisions, 11, 17, 22, 23, 24, 27, 31, 41, 46, 47, 94, 118, 119, 180, 182
defense(s), 53, 56
deficit, 59
definition, viii, 73
density, 127, 137, 142, 156, 181, 182
deposition, 96, 121, 179
deposits, 96

depression, 52
desire, vii, 62
desynchronization, 192
detection, x, 46, 133, 149, 150, 155, 156, 159, 191
detection techniques, 150
developmental process, 12
deviation, 12, 120
diet(s), 130, 165, 174, 182
dimorphism, xi, 26, 29, 193, 194, 195, 196, 201, 203, 206, 209, 210, 211, 212, 213
diploid, 120
direct costs, 39
direct observation, 121
discrimination, xi, 13, 36, 44, 49, 121, 186, 187, 188, 189, 190, 191, 192
discrimination learning, 191, 192
dispersion, 49, 182
distribution, x, 100, 113, 114, 118, 133, 134, 136, 142, 143, 149, 159, 160, 178
distribution function, 113
diurnal, 154, 160, 174
divergence, 43
diversification, 176, 183
diversity, 94, 95, 117, 213
diving, 153, 154, 155
DNA, 176
dogs, 75, 89
dominance, 17, 18, 19, 22, 24, 31, 32, 39, 40, 52, 53, 56, 70, 141, 179, 181, 206, 213
dominant animals, 172
dopaminergic, 63
Drosophila, 7, 10, 26, 29, 126, 188, 189, 190, 191, 192
drug abuse, 63
drug addiction, 63
drugs, 63
duration, 23, 62, 66, 108, 121, 124, 126, 134, 138, 147, 167, 175

E

ears, 128
earthworm(s), 165, 166, 175, 177
ecology, vii, ix, 1, 27, 28, 29, 94, 96, 118, 127, 130, 131, 133, 151, 152, 159, 181, 183
economic, 27
economics, 179
egg(s), 4, 14, 15, 16, 45, 94, 95, 96, 115, 117, 118, 119, 120, 121, 165
ejaculation, 19, 64, 71
elaboration, 206
election, 124, 182
electrophysiological, 190

emotionality, 64
emotion(s), vii, 61, 62, 64
enantiomers, 191
encapsulation, 123
encoding, 192
endocrine, 27, 68, 71
endogenous, 69
energy, 4, 14, 15, 124, 149, 151, 154, 156, 167, 170, 175, 179, 182
energy consumption, 156
England, 35
environment, ix, 2, 37, 39, 40, 41, 118, 133, 143, 146, 171, 176
environmental, vii, 12, 35, 36, 46, 47, 95, 118, 127, 135, 146, 210
environmental characteristics, 135
environmental conditions, 36, 47
equilibrium, 6
equipment, 47, 124
estimating, 58
ethology, vii, 71
Europe, 96, 114, 115, 116
European, ix, 53, 93, 95, 112, 113, 115, 118, 128, 160, 179
evening, 167
evidence, viii, xi, 8, 10, 12, 16, 17, 18, 19, 20, 22, 24, 25, 28, 31, 37, 39, 40, 41, 42, 43, 44, 45, 49, 53, 58, 61, 64, 73, 84, 85, 87, 88, 91, 100, 119, 121, 136, 141, 179, 191, 193, 194, 212, 213
evolution, ix, xi, 1, 2, 3, 4, 5, 6, 7, 8, 10, 12, 16, 17, 27, 28, 29, 45, 47, 48, 51, 52, 53, 55, 56, 77, 88, 89, 90, 93, 94, 95, 96, 123, 127, 128, 130, 131, 134, 136, 141, 152, 153, 157, 159, 165, 172, 174, 176, 178, 179, 180, 182, 183, 193, 194, 196, 209, 210, 211, 212, 213
evolutionary, vii, viii, x, 1, 2, 5, 6, 22, 23, 24, 26, 31, 36, 44, 45, 73, 77, 90, 94, 95, 124, 126, 127, 130, 163, 196, 210, 212, 213
evolutionary process, vii, 1, 2
exaggeration, 6, 9, 11
excitation, 62, 70
exercise, 37, 39, 45
exotic, 166
experimental condition, 46
experimental design, 66
exploitation, 95, 122
exponential, 113
exposure, 60, 61, 65, 66, 142, 191
extrinsic, 41
eye(s), 10, 74, 90, 195, 196, 211, 213

F

failure, 81, 149, 209
false negative, 149
false positive, 149
family, ix, 20, 93, 164, 172, 195
family members, 20
fat, 31, 170, 179
fauna, 165, 177
fear, 59, 66, 67, 70
feedback, 6
female preference, 6, 7, 8, 9, 10, 13, 18, 22, 23, 37, 43, 47
females, vii, viii, ix, xi, 1, 3, 4, 5, 6, 7, 9, 10, 11, 13, 14, 15, 16, 17, 18, 19, 20, 22, 24, 25, 27, 31, 35, 36, 37, 38, 39, 40, 41, 42, 43, 44, 45, 46, 47, 49, 52, 54, 56, 57, 58, 60, 61, 62, 63, 65, 67, 93, 95, 96, 107, 111, 113, 115, 117, 118, 119, 120, 121, 123, 129, 134, 136, 137, 140, 141, 142, 143, 144, 151, 156, 164, 165, 166, 170, 171, 172, 173, 174, 175, 193, 194, 195, 196, 201, 202, 209, 210, 213
fermentation, 186
fern, 166
fertility, 6, 14, 15, 16, 30, 41
fertilization, 14, 15, 31, 45
filters, 149
fish, 16, 50, 131, 199, 201, 213
fission, 77
fitness, vii, 2, 6, 9, 10, 12, 14, 18, 19, 28, 35, 36, 37, 38, 42, 43, 44, 45, 53, 94, 96, 115, 119, 120, 121, 123, 125, 144, 172
fixation, 43
flexibility, 120, 122
flies, ix, 93, 94, 95, 97, 103, 105, 106, 107, 108, 111, 113, 115, 116, 117, 118, 119, 120, 121, 122, 123, 124, 125, 129, 131, 186, 195, 197, 209, 210, 211, 213
flight, 124, 126
floating, 149
flora and fauna, 165
flow, 137
fluctuations, 26, 117
focusing, 62, 96
food, viii, x, 15, 16, 37, 38, 46, 63, 73, 75, 76, 77, 80, 88, 91, 135, 136, 163, 166, 167, 168, 169, 170, 171, 172, 173, 174, 175, 176, 177, 178, 179, 180, 181, 182, 192
food caches, ix, 74, 76
food hoarding, x, 163, 168, 169, 171, 173, 174
foraging, x, 16, 134, 136, 138, 140, 153, 163, 165, 170, 172, 174, 177, 182
founder effect, 28
Fourier, 156
France, 73, 112, 113, 185
free choice, 37, 39
freezing, 135
friendship, 19, 32
fruit flies, 10, 191
fruits, 165
functional analysis, 32
functional architecture, 186
funding, 190
Fur, 51, 53
fusion, 77

G

GABA, 63
gametes, 4
gauge, 149
gaze, viii, 23, 41, 73, 74, 75, 76, 77, 78, 79, 80, 81, 82, 83, 84, 85, 86, 87, 88, 89, 90, 91, 92
gene expression, 210
generalization, x, 185, 186, 187, 188, 189, 190, 191, 192
generation, 69
gene(s), 2, 6, 7, 9, 10, 11, 12, 13, 15, 21, 22, 31, 39, 40, 41, 42, 43, 44, 50, 51, 52, 53, 56, 190, 191, 194, 196, 210, 212
genetic, ix, xi, 7, 8, 9, 10, 12, 15, 16, 17, 20, 22, 26, 27, 30, 31, 36, 37, 39, 40, 42, 43, 44, 45, 46, 47, 53, 54, 56, 93, 124, 126, 127, 193, 194, 196, 209, 210, 211, 212, 213
genotype(s), 43, 44, 45, 54, 65, 68, 70, 71
Germany, 93, 133, 149, 155, 158
gestation, vii, 4, 35, 40
global warming, 135
glomerulus, 186
goal-directed behavior, 62
goals, 149, 150
gonad, viii, 57
Gore, 52
government, 147
Greece, 97, 98, 113, 122, 128
greek, 131
Greenland, 137, 160
groups, x, 11, 18, 27, 30, 49, 59, 77, 78, 87, 94, 108, 140, 142, 143, 146, 163, 172, 176, 181, 186, 194, 195, 201, 206
growth, ix, 12, 31, 93, 118, 119, 123
growth rate, 12
gryllid, ix, 93, 94
Guinea, 130
Gulf of Mexico, 157
gustatory, 188

H

habitat, ix, 108, 128, 133, 134, 135, 138, 141, 142, 144, 145, 151, 158, 182
habituation, 82, 186, 189, 192
harassment, 36, 38, 40, 50, 53, 54
harbour, 42, 51, 135, 140, 142, 152, 154, 157, 158, 160
Harvard, 26, 27, 32
harvesting, 182
Hawaii, 113, 114, 115, 120, 123, 152, 157
head, 74, 75, 76, 78, 81, 82, 117, 187, 196, 198, 201, 209
health, 20, 22, 38
hearing, 120, 122, 123, 128, 129, 130
heat, 41, 135
heat loss, 135
hedonic, 70
height, 166
hemisphere, 174
herbivores, 94
heritability, 6, 7, 8, 10, 28
heterozygosity, 43, 48
hip, 119
hippocampus, 178
hoarding, x, 163, 166, 168, 169, 170, 171, 172, 173, 174, 175, 176, 177, 179, 180, 181, 182
Homotrixa alleni, ix, 94, 96, 108, 109, 112, 113, 114, 117, 118, 125, 127, 131
honesty, 33, 54
hormone(s), xi, 58, 60, 69, 193, 210
host, ix, 62, 93, 94, 95, 96, 97, 98, 99, 100, 103, 104, 107, 108, 112, 113, 114, 115, 116, 117, 118, 119, 120, 121, 122, 123, 124, 125, 127, 128, 129, 130, 131
host population, 108, 123
human(s), viii, ix, x, 16, 17, 18, 19, 21, 22, 23, 24, 25, 29, 35, 67, 69, 73, 74, 75, 76, 78, 80, 81, 82, 86, 89, 90, 91, 149, 150, 163, 164, 165, 166, 167, 168, 178, 186, 189, 191
human animal, 74, 75, 86, 89
hunting, 96, 128, 182
hydrophone, 146, 147, 148, 153, 157, 159
hymenoptera, 95, 96
hypothesis, ix, 5, 6, 9, 10, 11, 19, 30, 32, 42, 45, 64, 77, 93, 142, 171, 172, 176, 181, 182, 209

I

ice packs, 135
ice-dominated, ix, 133, 135, 151
identification, 21, 155, 157, 168

identity, 56, 90, 117, 192
idiosyncratic, xi, 193, 210
Illinois, 159
images, 23, 48, 191
immune response, 123
immune system, 30, 43, 118
immunity, 130
immunocompetence, 27
immunosuppressive, 22
implementation, 25
in situ, 5, 6, 19
inbreeding, 30, 31, 43, 44, 52
incentives, 38, 64, 67, 70
incest, 19, 20
incidence, 99
incompatibility, 56
independent variable, 61
Indiana, 50
indication, ix, 5, 60, 61, 73, 84, 168
indicators, 9, 55, 117, 144
indices, 58
individual differences, 151
individuality, 137, 151, 153, 157
infancy, 90
infant care, 38
infant mortality, 31
infants, 20, 31, 38, 55, 90, 91
infection, 11, 12, 22, 28, 39, 54, 95, 97, 98, 101, 113, 119, 132
infectious diseases, 44
inferences, 24
infertile, 14
inheritance, 48
inhibition, 190
inhibitory, 186
injury, 170
insects, ix, 31, 37, 40, 41, 93, 94, 119, 123, 130, 131, 165, 167, 177
insemination, 54
insight, 18, 77, 87, 176
inspection, 202
instability, 127
institutions, 147
instruments, x, 133, 146, 149
integration, 25
intelligence, 90, 176, 183
intensity, 4, 6, 7, 21, 58, 59, 61, 65, 174, 192
interaction(s), 10, 18, 20, 23, 36, 43, 46, 47, 59, 60, 61, 64, 67, 69, 71, 75, 118, 127, 136, 140, 142, 151, 212
internet, 149, 201, 206
interneurons, 131, 186
interpretation, viii, 57, 68, 85, 88, 172

inter-sexual, viii, 3, 35, 41
interval, 84, 85, 128
intervention, 19
intrinsic, vii, 35, 41, 147
intrusions, 143
invertebrates, 126, 165, 166, 167, 175, 186
investigations, 151
investment, 4, 26, 32, 36, 52, 55, 95, 140, 154, 160
Ireland, 153
iridium, 148
isolated islands, x, 163
isolation, 29, 54, 81, 165

J

jackdaws, viii, 73, 74, 77, 81, 83, 84, 85, 87, 88, 91
Japan, 152
Japanese, 12, 19, 27, 30, 31, 32, 39, 55, 91
jays, x, 88, 91, 163, 174, 176, 180, 181, 182, 183
juveniles, 170

K

kin selection, 172
kinase, 191
krill, 152

L

lactating, 155, 156
lactation, 4, 134, 135, 136, 137, 138, 140, 141, 142, 156, 158
land, 135, 137, 138, 140, 141, 142, 154
language, 90
larva(e), 94, 95, 96, 97, 99, 100, 103, 107, 113, 115, 118, 119, 120, 125, 126, 127, 128, 131, 189, 190, 191, 192
larval, ix, 93, 94, 117, 120, 122, 125, 129
laser, 82, 83, 84
laser pointer, 82, 84
lattice, 156
law(s), 61, 192
lead, 6, 12, 16, 24, 43, 61, 62, 68, 83, 95, 145, 188, 189, 190, 194
learning, vii, viii, x, 57, 63, 71, 75, 88, 89, 91, 163, 173, 183, 185, 186, 187, 188, 189, 190, 191, 192
learning task, x, 185, 187, 188, 190
Lepidoptera, 94, 126
lichen, 166
life cycle, 128
life spans, vii, 35, 40
life style, 134

lifetime, 4, 5, 9, 39, 129
likelihood, 15, 16, 45, 136, 144, 172, 211
limitation, 95, 122
linear, 98, 100, 103, 117
listening, 139, 148
literature, vii, viii, 1, 3, 5, 8, 21, 24, 35, 37, 57, 59, 60, 61, 66, 174, 195
local area network, 148
localization, x, 117, 134, 147, 148, 153, 155, 159
location, 37, 117, 150, 153, 166, 172, 176
locomotion, 135
London, 26, 27, 28, 30, 49, 50, 51, 52, 53, 54, 56, 90, 126, 128, 129, 152, 153, 154, 155, 156, 157, 158, 176, 178, 212, 213
long distance, 143, 144
long period, 174, 175
longevity, 39, 174
long-term, x, 19, 43, 47, 52, 86, 133, 134, 144, 161, 174, 180, 181, 191, 192
long-term memory, 191, 192

M

maggots, ix, 94, 97, 99, 100, 107, 108, 113, 115, 118, 119, 121, 123, 190
maintenance, 7, 9, 19, 26, 139, 148, 157, 173
major decisions, 94, 119
major histocompatibility complex (MHC), 43, 44, 45, 50, 51, 53, 54, 56
maladaptive, xi, 6, 193, 194, 210, 211
male traits, viii, 5, 9, 11, 35, 41, 47
males, viii, ix, xi, 1, 2, 3, 4, 5, 7, 9, 10, 11, 12, 14, 15, 16, 17, 18, 19, 20, 22, 26, 27, 29, 31, 32, 36, 37, 38, 39, 40, 41, 42, 43, 44, 45, 46, 47, 48, 49, 50, 51, 52, 54, 56, 57, 58, 59, 60, 61, 62, 64, 65, 66, 67, 69, 70, 93, 95, 97, 98, 99, 100, 101, 103, 105, 106, 107, 108, 113, 115, 116, 117, 119, 120, 121, 123, 126, 129, 132, 137, 141, 142, 143, 144, 164, 165, 170, 171, 172, 173, 174, 175, 193, 194, 195, 196, 201, 203, 206, 209, 210
mammal, vii, x, 4, 35, 36, 37, 39, 40, 41, 42, 43, 44, 45, 46, 47, 134, 145, 146, 149, 150, 159, 165, 177
management, 175, 179, 182
manifold, 124
manipulation, 21, 24, 25, 31, 127
Markov, 150, 155
marsh, 181
masculinity, 25
masking, x, xi, 185, 186, 188, 189, 190
Massachusetts, 89, 127, 160, 212
matching-to-sample, 186, 192
mate choice, vii, 1, 2, 3, 4, 5, 9, 10, 11, 13, 14, 15, 16, 17, 18, 19, 20, 21, 22, 23, 24, 25, 27, 29, 31,

32, 33, 35, 36, 37, 38, 39, 40, 41, 42, 43, 44, 45, 46, 47, 48, 49, 50, 51, 52, 53, 54, 55, 56, 183, 206, 209, 213

mating behaviour, viii, 26, 29, 35, 48, 141, 142, 143, 144

mating preferences, vii, viii, 1, 3, 5, 7, 8, 9, 11, 14, 17, 18, 23, 28, 29, 30, 36, 37, 40, 43, 44, 45, 46, 47, 48, 51, 53, 56

mating songs, ix, 93, 95

mating strategies, 4, 36

maturation, 164, 177

maze learning, 71

measurement, 59, 60

measures, 46, 47, 61, 186, 188

mechanical, 130

median, 85

mediation, 63

Mediterranean, 96

melt(s), 135, 148

membranes, 95

memory, x, 89, 163, 173, 176, 183, 185, 190, 191, 192

men, 29, 70

mental representation, 75

metabolic rate, 128, 130, 167

Mexican, 176, 183

Mexico, 157

mice, viii, 41, 42, 43, 44, 45, 47, 50, 51, 52, 54, 55, 57, 58, 59, 60, 61, 63, 64, 67, 68, 69, 70, 71, 191

microbial, 174

micro-climate, 135

migration, 142, 144, 153, 154, 157

military, 147, 148, 159

milk, 156

MIT, 89

mites, 11

mitochondrial, 176

mitochondrial DNA, 176

model system, 188, 190

models, 1, 7, 9, 21, 30, 43, 46, 59, 78, 81, 82, 83, 84, 85, 86, 88, 89, 172, 179, 180, 186, 190, 211, 212, 213

modulation, 140, 191

mole, 49, 96, 108, 113, 116, 119, 122, 123, 124, 126, 127, 129

molecules, 213

Møller, 1, 5, 10, 11, 12, 13, 17, 26, 30

monkeys, 18, 20, 25, 26, 27, 28, 29, 30, 31, 39, 41, 42, 75, 88, 90

monograph, 65

monomorphic, xi, 193, 194, 196, 201, 202, 203, 206, 209, 210

morning, 167

morphology, 3, 154

mortality, 31, 54, 117, 119, 120, 123, 135, 177

mothers, 7, 8, 19, 28, 120, 137, 140, 154

motivation, viii, 2, 36, 43, 57, 58, 59, 60, 61, 62, 63, 64, 66, 68, 69, 70, 71, 78

motor activity, 66

movement, 165

multidimensional, 24

multiple interpretations, 65

muscles, 123

N

National Academy of Sciences, 29, 48, 52, 53, 56

National Park Service, 159

natural, vii, viii, 1, 2, 14, 17, 35, 36, 37, 39, 42, 46, 48, 62, 87, 94, 95, 121, 127, 145, 170, 173, 181

natural enemies, 94

natural environment, 46

natural food, 170

natural selection, vii, 1, 2, 95, 121

negative relation, 119

nematode, 11, 44, 192

Netherlands, 26, 155, 157, 193

network, 147, 148, 158

neuroanatomy, 191

neurobiological, 63, 186, 190

neurobiology, 71

neuroendocrine, 71

neurons, 186

neuroscience, 69

neurotransmitter, 65

New Jersey, 127, 181, 182

New World, 205, 206, 212

New York, 26, 27, 28, 30, 31, 32, 48, 55, 91, 126, 155, 159, 178, 181, 212

New Zealand, x, 163, 164, 165, 166, 167, 168, 171, 172, 173, 174, 175, 176, 177, 178, 181

nitric oxide (NO), 139, 190, 191, 192

nitric oxide synthase, 192

NOAA, 153, 156, 160

nodes, 196, 202, 211

noise, 145, 146, 149, 150, 151, 157, 158

non-human primates, 16, 17, 18, 19, 23, 24

North America, 11, 113, 114, 120, 122

North Atlantic, 156

Norway, 68

novelty, 209

null hypothesis, 209

nurse, 135, 144

nursing, viii, 57, 61, 136, 140, 154, 155

nutrient(s), 15, 135

nutrition, 15, 38

nuts, 174
nymphs, 113

O

obesity, 70
observational learning, 89, 173, 183
observations, viii, 19, 35, 61, 63, 79, 86, 121, 143, 147, 150, 152, 154, 156, 168, 170, 180
oceans, 147
odorants, 187, 191
Oedipus, 61
oil, 130
old age, 41
olfaction, 58, 65, 188
olfactory, x, xi, 64, 71, 137, 185, 186, 187, 188, 189, 190, 191, 192
olfactory bulb, 191
omission, 45
ontogenesis, 68
operant conditioning, 20
opioidergic, 63
Oregon, 159
organ, 122
organism, 2, 9, 211
organization, 77, 156
orientation, 70, 74, 75, 78, 87
ormiini fly, ix, 93
ornamentation, xi, 9, 10, 26, 193, 194, 201, 203, 206, 209, 210, 211, 212
ornaments, vii, xi, 1, 9, 10, 28, 30, 49, 193, 194, 209, 210, 211
outliers, 85
ovulation, 25, 46
ownership, 196
oxide, 190, 191, 192

P

P. thessalicus, ix, 93, 96, 97, 98, 99, 100, 101, 102, 103, 104, 105, 106, 107, 113, 117, 121, 123
Pacific, 143, 147, 152, 153, 155, 157, 159
pacing, 172, 173
Pan troglodytes, viii, 20, 39, 52, 55, 73, 75, 92
Papua New Guinea, 130
paradox, 29, 38
parameter, 66
parasite(s), ix, 5, 10, 11, 12, 13, 27, 28, 30, 33, 39, 42, 44, 50, 53, 93, 95, 130, 131
parasitic infection, 12, 28
parasitizing, ix, 93, 94, 113, 122

parasitoid, ix, 93, 94, 95, 96, 97, 99, 100, 102, 103, 107, 108, 110, 111, 113, 115, 116, 118, 119, 120, 121, 122, 124, 125, 126, 127, 128, 129, 130, 131, 132
parental care, 4, 18, 19, 20, 28, 32, 38, 181
parenting, 15
parents, 76, 80, 88
parids, x, 89, 163, 172, 178
Paris, 185
partition, viii, 20, 57, 58, 59, 60, 61, 62, 64, 66, 68, 70, 82, 83
passive, ix, x, 17, 133, 139, 145, 147, 154, 156
paternity, 17, 19, 36, 37, 38, 46, 48, 50, 54
pathogens, 22
pathology, 62
pathways, 190
perception(s), viii, 47, 49, 73, 75, 77, 186, 191, 192
PerenniAL Acoustic Observatory (PALAOA), x, 134, 139, 148, 150
performance, 50, 55, 118, 120, 124, 127, 129
permit, 16
personal, vii, 15
pests, 94
pharmacological, 58, 61, 65, 67, 70
pharmacological treatment, 67
pharmacology, 71
phenotype(s), viii, 35, 37, 38, 39, 41, 42, 43, 44, 45, 46, 48
phenotypic plasticity, 123, 125, 127
pheromones, 65, 71
phylogenetic, xi, 88, 126, 164, 176, 193, 194, 195, 196, 202, 210, 211, 212, 213
phylogenetic tree, 195, 196, 202
physiological constraints, xi, 193, 210, 211
physiological regulation, 68
physiology, 69, 118
pinnipeds, ix, 133, 134, 135, 136, 138, 139, 140, 141, 142, 144, 145, 146, 151, 154, 158, 159, 160
pituitary, 70, 71
plasma, 58, 70
plasticity, 46, 49, 123, 125, 127, 142, 151, 190, 213
play, ix, 2, 10, 17, 18, 42, 43, 94, 133, 136, 141, 154, 161, 173
pleasure, 62, 63, 70
pleiotropic effect, xi, 193, 210
plus-maze test, 58
Poecilimon mariannae, ix, 93, 95, 97, 98, 100, 101, 103, 104, 105, 106, 107, 109, 129
polar, ix, x, 133, 134, 135, 138, 139, 142, 144, 145, 146, 147, 148, 150, 151
polygenic, 29, 212
polymorphism, 7

population, vii, 1, 4, 7, 14, 37, 44, 53, 54, 97, 99, 118, 123, 127, 132, 137, 143, 156, 158, 159, 168, 180, 181
population density, 156
positive correlation, 39
positive emotions, 61, 62, 64
positive feedback, 6
positive reinforcement, 62, 63, 64
positive relationship, 10, 88
post-hoc analysis, 83
power, 147, 148
predators, ix, 46, 75, 76, 81, 88, 93, 131, 135, 165, 167, 174, 182
predictability, 136
prediction, 194, 209
preference, vii, x, 1, 2, 3, 5, 6, 7, 8, 9, 10, 11, 12, 14, 15, 16, 17, 18, 19, 20, 21, 22, 23, 24, 25, 26, 28, 30, 31, 35, 36, 37, 39, 40, 41, 43, 44, 45, 46, 47, 48, 52, 53, 54, 56, 63, 64, 71, 185, 186, 187, 201, 212
pregnancy, 58, 69
pressure, 4, 5, 6, 8, 9, 75, 127, 131
prevention, 91
primate(s), 1, 16, 17, 18, 19, 20, 21, 22, 23, 24, 25, 26, 27, 28, 29, 30, 31, 32, 37, 38, 40, 41, 45, 49, 50, 51, 55, 56, 75, 77, 91, 92, 175, 187, 191
probability, ix, 94, 171, 176
probe, 149
procedures, 89
production, 4, 12, 47, 95, 121, 144, 156, 174
productivity, 135
prolactin, 69, 71
promote, 89, 123, 210
propagation, 108, 141
protein(s), 45, 186, 191
protocol(s), 78, 85, 168, 190
psychiatry, 67
psychology, 71
psychophysics, 186
psychophysiological, 64
pulses, 143
puncture wounds, 166
pupa(e), ix, 94, 97, 100, 103, 105, 106, 107, 115, 116, 118
pupal, ix, 93, 103, 105, 108, 109, 119
pupation, 100, 120

Q

quartile, 85

R

race, 44
radiation, 164, 165
radio, 147, 148
random, 28, 41, 172
random mating, 28
range, ix, x, 24, 25, 86, 88, 93, 94, 108, 120, 122, 123, 128, 133, 143, 145, 153, 159, 182, 195
rat(s), 42, 49, 56, 60, 64, 68, 69, 70, 71, 89, 165, 181, 186, 192
raven(s), viii, x, 73, 74, 75, 76, 77, 78, 79, 81, 84, 85, 86, 87, 88, 89, 90, 91, 163
reactivity, 58, 65, 70
real time, 145, 147, 148, 153
reality, 45
reasoning, 3
reception, 157
receptors, 61, 65, 69, 71, 186, 189
recognition, 22, 27, 56, 59, 69, 94, 136, 137, 138, 140, 150, 151, 154, 155, 156, 160, 165, 186
recognition test, 138
recovery, 89, 181, 183
redness, 7
reduction, 2, 6, 11, 14, 22, 50, 66, 171, 212
refractory, 62, 63, 64, 95
regional, 120, 144
regression(s), 100, 101, 103, 196
regulation, 63, 65, 68, 71, 178, 179
reinforcement, 62, 63, 64
relationship(s), 4, 10, 16, 19, 22, 29, 39, 55, 87, 88, 94, 119, 124, 128, 129, 153, 158, 172, 173, 191, 212
relatives, 20, 201, 204
remote sensing, 145
reparation, 65
reproduction, ix, 2, 4, 36, 65, 68, 93, 125, 134, 168
reproductive activity, 54
reproductive success, vii, 4, 20, 28, 32, 35, 37, 38, 39, 40, 42, 43, 45, 46, 48, 50, 52, 55, 56, 59, 129, 173, 181, 183
reptile, 165
researchers, 39, 124, 175, 194
reserves, 15, 134, 151, 179
resistance, 10, 11, 12, 22, 44, 53
resolution, 78
resource availability, 168
resources, 15, 16, 36, 37, 40, 94, 96, 119, 123, 131, 147, 169, 174, 181
respiratory, 123
responsiveness, 74, 87
rhythm, 183, 190
rickets, 96, 108, 113, 119, 122, 123, 126, 127

risk, 14, 36, 38, 44, 46, 120, 135, 170, 179
roars, vii, 35, 43, 46, 49, 54
robins, x, 163, 164, 165, 166, 167, 168, 170, 171, 172, 173, 174, 175, 176, 177, 178, 181
rodents, 180, 187
rooks, viii, 73, 74, 76, 77, 78, 79, 80, 81, 84, 85, 87, 88
Royal Society, 25, 28, 49, 50, 51, 52, 53, 54, 56, 90, 126, 128, 129, 154, 176, 212, 213
ruminant, 53
Russia, 57
Russian, 57, 68, 69, 70, 71, 155

S

saliva, 166
salt, 191
sample, 107, 146, 186, 192
sampling, 36, 47, 100, 124, 139, 146, 147
satellite, 113, 117, 123, 146, 147, 148, 149, 150, 158
satisfaction, 62
scarcity, 174, 175
scatter, 172, 179, 180, 181
science, 74, 192
scores, 206
scrotal, 41
scroungers, ix, 74
sea ice, 134, 135, 145, 149, 152
seals, 42, 43, 48, 50, 51, 53, 134, 135, 136, 137, 140, 141, 142, 143, 144, 145, 149, 151, 152, 153, 154, 155, 156, 157, 158, 159, 160
search(es), 31, 62, 63, 67, 165, 172, 195
searching, 6, 36, 38, 47, 108, 118
seasonality, 96, 108, 178
secondary sexual characteristics, 7, 9, 11
secretion, 69
security, 171
sedentary, 137
seed(s), 89, 119, 123, 125, 165, 169, 171, 174, 176, 179, 180, 182, 183
selecting, vii, 1, 5, 7, 9, 38, 40
selectivity, 4
sensing, 145
sensitivity, x, 64, 149, 185, 187, 210
sensitization, 71, 190
sensors, 154
separation, 78, 134, 136, 154, 157, 165
septum, 143
series, 10, 11, 18, 20, 142, 168, 188
serotonin, 61, 65, 68, 69, 192
serum, 38, 71, 152
severe stress, 67

sex, vii, xi, 1, 2, 3, 4, 11, 23, 26, 27, 28, 31, 46, 49, 69, 87, 95, 103, 107, 118, 120, 123, 131, 157, 164, 171, 193, 194, 209, 212
sex ratio, 28, 69, 107, 120, 131
sexual activity, 19
sexual behavior, viii, 57, 58, 60, 61, 62, 63, 64, 68, 69, 71
sexual behaviour, 27, 28
sexual contact, 64
sexual dimorphism, xi, 26, 29, 193, 194, 195, 196, 201, 206, 209, 210, 211, 212, 213
sexual harassment, 50, 54
sexual motivation, viii, 57, 58, 59, 60, 61, 62, 63, 64, 65, 66, 67, 68
sexual orientation, 70
sexual pathology, 62
sexual reproduction, 168
sexually transmitted disease, 22
shade, 180
shape, 22, 23, 151
sharing, 189
sheep, 39, 44, 54
shelter, 108, 123, 135
shipping, 145
short period, 166, 175
shortage, 174
short-range, 143
shrikes, x, 26, 163, 175, 183
shrubs, 97
sibling(s), 8, 81, 85, 86, 87, 88, 91
sigmoid, 189
signalling, ix, 10, 12, 16, 26, 43, 95, 122, 126, 133, 136, 175, 190
signals, ix, 7, 9, 22, 27, 70, 93, 117, 131, 136, 141, 144, 147, 155, 186, 187, 190
signs, 201
similarity, x, 43, 44, 185, 187, 188, 189, 190, 191
sites, 157, 166, 170, 172, 173, 174, 176
skills, viii, ix, 73, 74, 75, 76, 77, 78, 79, 87, 88, 89
skin, 21, 22, 26
Slovenia, 122
small mammals, 40, 165
sociability, 58
social behavior, 25, 32, 67, 68, 70, 159
social cognition, 180
social conflicts, 68
social context, 90, 172, 179, 180
social environment, 39, 171, 176
social factors, vii, x, 35, 46, 163
social group, 19, 20, 25, 36, 181
social learning, x, 91, 163
social organization, 77
social situations, 171

social status, 50, 58, 59, 67
social stress, 68
social systems, 17
social units, 77
software, 149, 151
solar, 149
songbirds, 164, 168, 170, 174
sounds, x, 55, 125, 133, 134, 140, 141, 142, 143,
    145, 146, 149, 151, 152, 154, 156, 159, 161
South Africa, 182
South America, 38, 49, 157
Spain, 160
spatial location, 172
spatial memory, x, 89, 163, 176, 183
specialization, 127
speciation, 29, 212
specificity, ix, 93, 192
speculation, 168
speech, 150, 155
speed, 187
sperm, 4, 6, 14, 30, 39, 40, 41, 44, 45, 156, 157
stability, 136, 137
stages, 122
standardization, 67
statistical analysis, 79
stereotypical, 17, 141
steroids, 69
stimulus, viii, 47, 57, 58, 62, 67, 87, 175, 187
stock, 7
storage, 89, 139, 146, 147, 148, 149, 166, 167, 169,
    170, 171, 174, 180, 183
strain, 58, 59, 61, 64, 65
strategies, viii, x, 3, 4, 5, 26, 27, 35, 36, 46, 47, 48,
    50, 55, 56, 91, 95, 96, 117, 118, 125, 126, 127,
    134, 136, 141, 142, 143, 151, 154, 155, 160, 172,
    173, 176, 179, 180, 181, 185, 186
strategy use, 188
strength, 4, 47
stress, 42, 61, 66, 67, 68, 69, 70
stressors, 12
subgroups, 206
substrates, 69, 134
suffering, 95
sugar, 187, 191
summer, 53, 87, 113, 135, 164, 170, 174
superiority, 18, 19
superparasitism, ix, 94, 96, 100, 102, 108, 113, 119,
    121, 129, 130
supply, 14, 135, 147, 148, 174
suppression, 130, 181
surplus, 139, 166, 178, 182
surveillance, 147

survival, ix, 2, 4, 5, 6, 9, 12, 15, 16, 24, 31, 36, 42,
    53, 93, 94, 95, 106, 119, 125, 127, 177, 179, 181,
    193
survival rate, 106
surviving, 44, 96, 100
survivors, 120
susceptibility, 159
swarms, 120
Sweden, 30
switching, 124, 165, 173
Switzerland, 96
symbolic, vii
symmetry, 12, 22, 23, 24, 25, 27, 29
synchronous, 135
synthesis, 15
systems, x, 17, 22, 38, 39, 41, 45, 63, 65, 94, 96,
    108, 113, 115, 117, 122, 124, 125, 127, 134, 139,
    141, 146, 147, 148, 151, 155, 156, 157, 159, 163,
    166, 173, 174, 186

T

tactics, 4, 36, 38, 39, 45, 50, 108, 141, 142, 143, 144,
    151, 160
taxa, xi, 5, 20, 75, 77, 193, 194, 195, 202, 206, 209,
    210
taxonomic, ix, 93, 176
technological, 47
technology, 145, 151, 159
temperature, 56, 134, 135, 179
temporal, viii, 73, 78, 127, 136, 149, 156, 160
temporal distribution, 160
tenure, 20, 41, 50
territoriality, x, 28, 163
territory, 16, 37, 38, 46, 66, 142, 165, 170, 173, 175
testes, 68, 69
testosterone, viii, 22, 38, 57, 58, 59, 60, 61, 62, 64,
    65, 66, 68, 69, 70, 71, 152, 210
theft, 89, 167, 171, 172, 173, 175, 176, 181, 182
theoretical, vii, 1, 2, 3, 6, 7, 10, 17, 41, 45, 46, 89,
    95, 131, 179, 180, 194
theory, viii, x, 2, 3, 10, 11, 27, 29, 32, 36, 40, 42, 43,
    48, 57, 62, 63, 71, 91, 115, 119, 123, 163, 171,
    172, 182, 189
Therobia leonidei, ix, 93, 95, 96, 97, 98, 99, 100,
    101, 104, 108, 109, 112, 113, 114, 115, 118, 120,
    124, 128
thinking, 194
thorax, 117, 123, 196
threat, 19, 43, 166, 174
three-dimensional, 154
threshold(s), 32, 121, 123, 210, 211
threshold level, 210

throat, 19, 202
time periods, 145
timing, 46, 134, 136, 144
tits, x, 30, 89, 163, 169, 174, 178, 180, 181, 182
Tokyo, 152
tomato, 126
tracking, 153
trade-off, 149, 179
traffic, 145
traits, vii, ix, xi, 2, 3, 4, 5, 7, 8, 9, 10, 11, 12, 16, 18, 20, 21, 22, 24, 29, 35, 37, 40, 41, 43, 46, 47, 50, 93, 94, 95, 118, 123, 193, 194, 195, 196, 209, 212, 213
transformations, 21
transition(s), 141, 195, 201, 210, 211
transmission, 22, 29, 147, 148, 149
transport, 181
trees, 174, 195
trend, 99
trial, 78, 81, 167
Turkey, 96

U

UK, 1, 73, 178
Ukraine, 96
ultrasound, 124, 156
uncertainty, 119, 120, 124, 125
unconditioned, 64
underwater, ix, 42, 133, 134, 139, 140, 141, 142, 143, 144, 147, 149, 151, 152, 153, 154, 156, 157, 158, 159, 160, 161
underwater vehicles, 139, 147, 152
United States, 48, 52, 53, 56, 113
univariate, 54
updating, 173
urine, 42

V

values, 47, 111
variability, 27, 36
variable(s), 17, 61, 100, 118, 136, 137, 146, 149, 187, 206
variance, 4, 134
variation, 2, 7, 12, 20, 26, 30, 32, 36, 41, 44, 47, 49, 50, 51, 53, 94, 96, 97, 98, 100, 118, 119, 120, 122, 123, 129, 132, 136, 137, 144, 153, 154, 155, 156, 157, 159, 160, 164, 179, 180, 183, 212
vegetation, 16, 108, 165, 166
vehicles, 139, 147, 152
Vermont, 87
vertebrates, 88, 186
vessels, 145, 146, 147
vibration, 130
Victoria, 163, 168, 177, 178, 212
video, 82
violent, 165
viruses, 118
visible, 76, 83, 97, 166
visual, viii, 23, 41, 48, 67, 73, 74, 75, 76, 78, 80, 81, 83, 84, 87, 90, 91, 92, 137, 145, 146, 147, 175
visual attention, 74, 90, 91
visual field, 74
visual perception, viii, 73, 75
vocalizations, 5, 18, 51, 53, 134, 137, 140, 141, 142, 143, 144, 146, 149, 150, 151, 152, 153, 154, 155, 156, 157, 158, 159, 160
voice, 152
vomeronasal, 71

W

walking, 117, 129, 168
war years, 69
water, 16, 135, 136, 137, 140, 141, 142, 143, 147
wavelet, 156
weight ratio, ix, 94
West Africa, 50
wind, 135, 149
winter, 53, 91, 135, 158, 167, 169, 170, 171, 173, 174, 175, 179, 181, 182
WLAN, 148, 149
women, 29
woodpeckers, x, 163, 172, 174, 182
worms, 167, 171

Y

yield, 39, 43

Z

zygote, 4